Managing the Nuclear Fuel Cycle: Policy Implications of Expanding Global Access to Nuclear Power

Mary Beth Nikitin, Coordinator
Specialist in Nonproliferation

Anthony Andrews
Specialist in Energy and Defense Policy

Mark Holt
Specialist in Energy Policy

October 19, 2012

Congressional Research Service
7-5700
www.crs.gov
RL34234

CRS Report for Congress
Prepared for Members and Committees of Congress

Summary

After several decades of widespread stagnation, nuclear power has attracted renewed interest in recent years. New license applications for 30 reactors have been announced in the United States, and another 548 are under construction, planned, or proposed around the world. In the United States, interest appears driven, in part, by tax credits, loan guarantees, and other incentives in the 2005 Energy Policy Act, as well as by concerns about carbon emissions from competing fossil fuel technologies.

A major concern about the global expansion of nuclear power is the potential spread of nuclear fuel cycle technology—particularly uranium enrichment and spent fuel reprocessing—that could be used for nuclear weapons. Despite 30 years of effort to limit access to uranium enrichment, several undeterred states pursued clandestine nuclear programs, the A.Q. Khan black market network's sales to Iran and North Korea representing the most egregious examples. However, concern over the spread of enrichment and reprocessing technologies may be offset by support for nuclear power as a cleaner and more secure alternative to fossil fuels. The Obama Administration has expressed optimism that advanced nuclear technologies being developed by the Department of Energy may offer proliferation resistance. The Administration has also pursued international incentives and agreements intended to minimize the spread of fuel cycle facilities.

Proposals offering countries access to nuclear power and thus the fuel cycle have ranged from requesting formal commitments by these countries to forswear sensitive enrichment and reprocessing technology, to a de facto approach in which states would not operate fuel cycle facilities but make no explicit commitments, to no restrictions at all. Countries joining the U.S.-led Global Nuclear Energy Partnership (GNEP), now the International Framework for Nuclear Energy Cooperation (IFNEC), signed a statement of principles that represented a shift in U.S. policy by not requiring participants to forgo domestic fuel cycle programs. Whether developing states will find existing proposals attractive enough to forgo what they see as their "inalienable" right to develop nuclear technology for peaceful purposes remains to be seen.

GNEP was transformed into IFNEC under the Obama Administration and has continued as an international fuel cycle forum, but the Bush Administration's plans for constructing nuclear fuel reprocessing and recycling facilities in the United States have been halted. Instead, the Obama Administration is supporting fundamental research on a variety of potential waste management technologies. Other ideas addressing the potential global expansion of nuclear fuel cycle facilities include placing all enrichment and reprocessing facilities under multinational control, developing new nuclear technologies that would not produce weapons-usable fissile material, and developing a multinational waste management system. Various systems of international fuel supply guarantees, multilateral uranium enrichment centers, and nuclear fuel reserves have also been proposed.

Congress will have a considerable role in at least four areas of oversight related to fuel cycle proposals. The first is providing funding and oversight of U.S. domestic programs related to expanding nuclear energy in the United States. The second area is policy direction and/or funding for international measures to assure supply. A third set of policy issues may arise in the context of U.S. participation in IFNEC or related initiatives. A fourth area in which Congress plays a key role is in the approval of nuclear cooperation agreements. Significant interest in these issues is expected to continue in the 112th Congress.

Contents

Introduction 1
Renewed Interest in Nuclear Power Expansion 2
 Worldwide Nuclear Power Status 6
 Nuclear Fuel Services Market 7
  Yellowcake 8
  Conversion 10
  Enrichment 11
  Fuel Fabrication 14
 Final Stages of the Fuel Cycle 15
 Waste Disposal and Energy Security 16
Proposals on the Fuel Cycle 17
 Comprehensive Proposals 18
  El Baradei Proposal (2003) 18
  President Bush's 2004 Proposal 19
  Russia's "Global Nuclear Power Infrastructure" 19
 Assurance of Fuel Supply: Supplier Guarantees 20
  Six Country Concept 20
  World Nuclear Association 21
  Japan's IAEA Standby Arrangements System 21
  UK "Nuclear Fuel Assurance (NFA)" 22
 Assurance of Fuel Supply: Fuel Reserves 22
  U.S. LEU Fuel Reserve, the "American Assured Fuel Supply" (AFS) 22
  IAEA LEU Fuel Bank 23
  Russian LEU Fuel Reserve, Angarsk 25
 Assurance of Supply: Enrichment Services 26
  International Uranium Enrichment Center (IUEC), Angarsk, Russia 26
  Germany's Multilateral Enrichment Sanctuary Project (MESP) 27
 Back-End Fuel Cycle Proposals 27
  International Framework for Nuclear Energy Cooperation (IFNEC) 27
 Supply-Side Approaches 30
  Nuclear Suppliers Group 30
  Group of Eight Nations (G-8) 32
Comparison of Proposals 33
Prospects for Implementing Fuel Assurance Mechanisms 36
Issues for Congress 37

Figures

Figure 1. The Conceptual Nuclear Fuel Cycle 8
Figure 2. World Wide Nuclear Power Plants Operating, Under Construction, and Planned 39

Tables

Table 1. Multilateral Nuclear Fuel Assurance Proposals 1
Table 2. Commercial UF$_6$ Conversion Facilities 11
Table 3. Operating Commercial Uranium Enrichment Facilities 12
Table 4. Comparison of Major Proposals on Nuclear Fuel Services and Supply
  Assurances 34

Contacts

Author Contact Information 40
Acknowledgments 40

Introduction

This report is intended to provide Members and congressional staff with the background needed to understand the debate over proposed strategies to redesign the global nuclear fuel cycle. It begins with a look at the motivating factors underlying the resurgent interest in nuclear power in some parts of the world, the nuclear power industry's current state of affairs, and the interdependence of the various stages of the nuclear fuel cycle.

After languishing for several decades, nuclear power in the United States appears poised for new growth, with license applications announced for up to 30 new commercial reactors. Two new U.S. uranium enrichment plants are currently under development in anticipation of an increased demand for nuclear fuel, and two others have been proposed. However, no U.S. facilities are currently planned for reprocessing spent nuclear fuel—the separation of uranium and plutonium to make new fuel. Other countries provide commercial reprocessing services and, with several notable exceptions, have kept their commercial and weapons fuel cycles separate.

Renewed interest in expanding the role of nuclear power in meeting world energy demand, particularly in countries considering their first nuclear power plants, has also led to increased concerns about limiting the spread of nuclear weapons-relevant technology. Most of this concern focuses on the nuclear fuel cycle, which includes facilities that can be used to make fuel for nuclear reactors or materials for weapons. Concerns about the nuclear fuel cycle have been increased by several high-profile cases of subversion of ostensibly commercial uranium enrichment and reprocessing technologies for military purposes. In 2003 and 2004, it became evident that Pakistani nuclear scientist A.Q. Khan had sold sensitive technology and equipment related to uranium enrichment to states such as Libya, Iran, and North Korea.

Leaders of the international nuclear nonproliferation regime have suggested ways of reining in the diffusion of such inherently dual-use technology, primarily through the creation of incentives not to enrich uranium or separate plutonium. Because a major justification for building enrichment or reprocessing facilities as part of a nuclear power program is to ensure fuel supplies for a nation's nuclear power plants, proposals to discourage these facilities in new states have focused on alternative ways to guarantee supplies of nuclear fuel.

Table 1. Multilateral Nuclear Fuel Assurance Proposals

Year	Agency	Proposal
2003	IAEA	Would establish internationally owned fuel cycle centers.
2004	United States	Would keep uranium enrichment and plutonium reprocessing in the hands of current technology holders, while providing fuel guarantees to those who abandon the option.
2005	IAEA	Explored a variety of options to address front end and back end problems and their attractiveness to different groups of states, and surveyed past proposals.
2005	Russian Federation	Would establish international fuel cycle centers.
2006	United States	U.S. Global Nuclear Energy Partnership originally proposed that certain recognized fuel cycle countries would ensure reliable supply to the rest of the world in return for commitments to renounce enrichment and reprocessing; also proposed solutions for recycling of spent fuel and storage issues.

Year	Agency	Proposal
2006	U.S., U.K., Russia, France, Germany, and Netherlands	Six Country Concept would establish reliable access to nuclear fuel.
2006	Nuclear Threat Initiative	Promised $50 million for an international nuclear fuel bank under IAEA supervision provided another $100 million donated within two years and IAEA organizes implementation.
2007	United States	Revised GNEP would promote an international nuclear fuel supply framework (without explicit renunciation of fuel technology) to reduce proliferation risk and a closed fuel cycle featuring recycling techniques that do not separate plutonium.

Most of the proposals are not new, but rather variations of those developed 30 or more years ago.[1] In the 1970s, efforts to limit or manage the spread of nuclear fuel cycle technologies for nonproliferation reasons foundered for technical and political reasons, but many states were nevertheless deterred from enrichment and reprocessing simply by the high technical and financial costs of developing sensitive nuclear technologies, as well as by a slump in the nuclear market. Several developments may now make efforts to limit access to the nuclear fuel cycle more feasible and timely: a growing concern about the spread of enrichment technology (specifically via the A.Q. Khan black market network, as well as Iran's enrichment program); a growing consensus that the world must seek alternatives to polluting fossil fuels; and optimism about new nuclear technologies that may offer more proliferation-resistant systems. Central to the debate is developing proposals attractive enough to compel states to forgo what they see as their inalienable right to develop nuclear technology for peaceful purposes. At the same time, there is debate on how to improve the International Atomic Energy Agency (IAEA) safeguards system and its means of detecting diversion of nuclear material to a weapons program in the face of a worldwide nuclear power expansion.

Renewed Interest in Nuclear Power Expansion[2]

Commercializing nuclear power has proved far more challenging than supporters of the technology had first envisioned. After the first wave of commercial reactor orders in the 1960s and 1970s, world nuclear capacity reached about 200 gigawatts during the 1980s, but as confidence in nuclear power safety declined after accidents at Three Mile Island and Chernobyl, the rate of further capacity additions fell more than 75% during the following decade.[3] Today, nuclear power plants have a total capacity of about 372 gigawatts—providing about 14% of the world's electricity generation.[4] Though a significant amount, it is far less than was generally projected 50 years ago. High construction and operating costs, safety problems and accidents, and controversy over nuclear waste disposal slowed the worldwide growth of nuclear power.

With uranium once considered a scarce resource, reprocessing and fast breeder technology was developed as a means of capturing the large amounts of potential energy remaining in spent

---

[1] See timeline of fuel cycle proposals, available at http://www.iaea.org/NewsCenter/Focus/FuelCycle/key_events.shtml.

[2] This section was prepared by Mark Holt, Specialist in Energy Policy, and Anthony Andrews, Specialist in Energy Policy, in the Resources, Science, and Industry Division, Congressional Research Service.

[3] International Energy Agency, *World Energy Outlook 2011*, p. 450.

[4] Energy Information Administration, *International Energy Outlook 2011*, September 19, 2011, Figure 75, http://www.eia.gov/forecasts/ieo/electricity.cfm.

nuclear fuel after it had been discharged from a reactor. In the 1980s, as the economics of advanced nuclear technology became questionable with declining fossil fuel prices and increased uranium supplies, national programs to develop fast breeder reactors came nearly to a standstill. Moreover, the plutonium fuel produced by breeder reactors drew strong opposition over its potential use in nuclear weapons.

In the past few years, the original promises of nuclear power attracted renewed interest around the world and led to predictions of a "nuclear renaissance." The March 2011 accident at Japan's Fukushima Dai-ichi nuclear plant has prompted some countries, notably Japan, to reverse their nuclear expansion plans, but others, such as China, have continued building new reactors. Therefore, while safety is clearly a major concern, several other key considerations are also driving current nuclear energy policy.

Volatile prices for oil and natural gas are a fundamental factor in national energy policymaking. Average world prices for a barrel of oil rose from below $10 at the beginning of 1999 to above $130 in mid-2008. They then declined to around $50 in early 2009 and rose to around $100 through mid-2012.[5] U.S. natural gas prices have been similarly volatile, although falling sharply in 2012 with increased production from shale formations.[6] To reduce their vulnerability to oil and gas price swings, national governments are searching for alternative energy sources, often including nuclear power. However, only 21% of the world's electricity generation is fueled by natural gas and 5% by oil,[7] so nuclear power's ability to directly substitute for oil and gas is limited, at least in the near term.

For nuclear power to have a significant impact on oil demand, long-term changes in energy-use patterns would have to take place, particularly in the transportation sector. One possibility is that nuclear power plants could be used to produce hydrogen, which could provide energy for fuel-cell vehicles. The U.S. Department of Energy has been developing processes that could produce hydrogen in a high-temperature reactor, an effort that has continued under the Obama Administration. Another possibility is the commercialization of all-electric or plug-in hybrid vehicles that could be recharged with nuclear-generated electricity. But even if such technologies were to be successfully developed, it would take many years for the new vehicles and, in the case of hydrogen, fuel delivery infrastructure to have a significant energy impact.

Government policies aside, volatile oil and gas prices are having a significant effect on projections of nuclear power's economic viability. In the United States, natural gas has been the overwhelming fuel of choice for new electrical generation capacity since the early 1990s, although coal-fired plants still generate about 40% of U.S. and world electricity.[8] Because fuel costs constitute a relatively small percentage of nuclear power costs, higher natural gas prices could make new nuclear power plants economically competitive, despite higher uranium prices.[9]

Growing worldwide concern about greenhouse gas emissions, particularly carbon dioxide from fossil fuels, has renewed attention to nuclear power's lack of direct CO_2 emissions. President

---

[5] EIA, "F.O.B. Costs of Imported Crude Oil by Area," http://www.eia.gov/dnav/pet/pet_pri_imc1_k_m htm.
[6]. EIA, "U.S. Natural Gas Wellhead Price," http://www.eia.gov/dnav/ng/hist/n9190us3m htm.
[7] World Energy Outlook, *op. cit.*, p. 546.
[8] World Energy Outlook, *op.cit.*, p. 546; EIA, *Electric Power Monthly*, Table 1.1, http://www.eia.gov/electricity/data.cfm#generation.
[9] CRS Report RL33442, *Nuclear Power: Outlook for New U.S. Reactors*, by Larry Parker and Mark Holt.

Obama's 2011 State of the Union address explicitly included nuclear power as part of the nation's "clean energy" strategy. Policies to reduce greenhouse gas (GHG) emissions may also indirectly encourage nuclear power expansion by increasing the cost of electricity from new fossil-fuel-fired power plants above that of nuclear power plants.

Some support for using nuclear power to reduce GHG emissions has emerged in academic and think-tank circles. As stated by the Massachusetts Institute of Technology in its major study *The Future of Nuclear Power*: "Our position is that the prospect of global climate change from greenhouse gas emissions and the adverse consequences that flow from these emissions is the principal justification for government support of the nuclear energy option."[10] But environmental groups generally contend that the nuclear accident, waste, and weapons proliferation risks posed by nuclear power outweigh any GHG benefits. The large construction expenditures required by commercial reactors, they contend, would yield greater GHG reductions if used for energy efficiency and renewable generation. Finally, they note that nuclear power, while not directly emitting greenhouse gases, produces indirect emissions through the nuclear fuel cycle and during plant construction.

Another key factor behind the renewed interest in nuclear power is improved performance of existing reactors. U.S. commercial reactors have generated electricity at an average of around 90% of their total capacity for the past decade, after averaging around 75% in the mid-1990s and around 60% in the mid-1980s.[11] Worldwide performance has seen similar improvement.[12] The improved operation of nuclear power plants has helped drive down the cost of nuclear-generated electricity. Average U.S. reactor operations and maintenance costs (including fuel but excluding capital costs) dropped steadily from a high of about 3.5 cents/kilowatt-hour (kwh) in 1987 to below 2 cents/kwh in 2001 (in 2001 dollars).[13] U.S. nuclear plant operating costs in 2011 averaged about 2.2 cents/kwh in current dollars.[14]

Nuclear interest has been further increased in the United States by incentives in the Energy Policy Act of 2005 (P.L. 109-58). The law provides a nuclear energy production tax credit for up to 6,000 megawatts of new nuclear capacity, compensation for regulatory delays for the first six new reactors, and federal loan guarantees for nuclear power and other advanced energy technologies. Under certain baseline assumptions, the tax credits and loan guarantees could determine whether new U.S. nuclear plants would be economically viable.[15]

U.S. electric utilities and other companies during the past five years have announced plans to submit license applications to the Nuclear Regulatory Commission (NRC) for about 30 new commercial reactors. NRC has issued "early site permits"—which resolve site-related issues for possible future reactor construction—at locations in Illinois, Mississippi, Virginia, and Georgia. The Tennessee Valley Authority in 2007 restarted construction of its long-delayed Watts Bar 2 reactor, which had been ordered in 1970. NRC in early 2012 issued combined construction and operating licenses for four new reactors in Georgia and South Carolina, allowing full construction

---

[10] Interdisciplinary MIT Study, *The Future of Nuclear Power*, Massachusetts Institute of Technology, 2003, p. 79.

[11] EIA, *Nuclear Power Plant Operations, 1957-2009*, http://www.eia.doe.gov/aer/txt/ptb0902.html.

[12] *Nuclear Engineering International*, November 2005, p. 37.

[13] Uranium Information Centre, *The Economics of Nuclear Power*, Briefing Paper 8, January 2006, p. 3.

[14] Nuclear Energy Institute, "Costs: Fuel, Operation and Waste Disposal," http://www.nei.org/resourcesandstats/nuclear_statistics/costs.

[15] CRS Report RL34746, *Power Plants: Characteristics and Costs*, by Stan Mark Kaplan.

to proceed. The status of the other planned new U.S. reactors appears to be more tentative, however. (For details on U.S. nuclear construction plans, see CRS Report RL33558, *Nuclear Energy Policy*, by Mark Holt.)

New reactors are on order elsewhere in the world, and several non-nuclear countries have announced that they are considering the nuclear option. As **Figure 2** shows, the vast majority of reactors currently under construction are in Asia, with only a handful in the rest of the world, primarily Russia.

Despite the recent positive developments for nuclear power, much uncertainty still remains about its prospects. Construction costs for new nuclear power plants—which were probably the dominant factor in halting the first round of nuclear expansion—continue to loom as a potential insurmountable obstacle to renewed nuclear power growth. Average U.S. nuclear plant construction costs more than doubled from 1971 to 1978, according to the Office of Technology Assessment, and nearly doubled again by the mid-1980s, not including interest accrued during construction.[16] Including interest, many U.S. nuclear plants proved to be grossly uneconomic, often with capital costs totaling more than $3,000 per kilowatt of capacity in 2000 dollars,[17] and relying on the utility regulatory system to recover their costs.

Major reactor vendors, such as General Electric and Westinghouse, have contended that new designs and construction methods will cut costs of future reactors considerably. However, the Energy Information Administration (EIA) estimates that new U.S. nuclear plants would cost $5,300 per kilowatt, excluding interest, making them potentially more expensive than the previous generation of reactors. EIA's estimates of the capital costs of several major competing power generation technologies, particularly coal and wind, have also risen sharply.[18]

Many other important factors in the future of nuclear power are similarly uncertain. Prices of competing fuels, especially natural gas, have been volatile in the recent past. If fossil fuel prices become depressed for a sustained period, as in the late 1980s through the 1990s, support for nuclear power as an alternative energy source could again be undermined. Disposal of high-level nuclear waste, which reprocessing or recycling is intended to address, will continue to generate controversy as governments attempt to develop permanent underground repositories—none of which are yet operating.

As noted above, the Fukushima Dai-ichi accident has clearly dimmed overall public support for nuclear power around the world, as did the earlier accidents at Three Mile Island and Chernobyl. New reactor designs are intended to be less vulnerable to the loss of backup power that crippled the Fukushima plant. But whether such safety features will significantly mitigate the concerns raised by the Fukushima accident remains to be seen.

---

[16] Office of Technology Assessment, *Nuclear Power in an Age of Uncertainty*, OTA-E-216, February 1984, p. 59.

[17] Jan Willem Storm van Leeuwen and Philip Smith, *Nuclear Energy, the Energy Balance*, July 31, 2005, Chapter 3, p. 2.

[18] Energy Information Administration, *Updated Capital Cost Estimates for Electricity Generation Plants*, November 2010, p. 8, http://www.eia.doe.gov/oiaf/beck_plantcosts/pdf/updatedplantcosts.pdf.

Worldwide Nuclear Power Status

Operating commercial nuclear reactors around the world currently total 433, which have an installed electric generating capacity of 372 gigawatts.[19] About 83% of world nuclear capacity is in member nations of the Organization for Economic Cooperation and Development (OECD), while slightly more than 13% is in Russia and other former nations of the Soviet bloc. The remainder, less than 5%, is in developing countries such as China and India. Nuclear power supplied 21.5% of electricity generated in OECD countries and 4.7% in non-OECD countries in 2009.[20]

Unlike the United States, where only one long-deferred reactor is currently under construction, much of the rest of the world has continued building nuclear plants, although at a modest pace. Since 1996, about 50 commercial reactors have started up, an average of about four per year. About 40 reactors were permanently closed during that period, although many of them were smaller than the newly started reactors.[21]

Current reactor construction is dominated by Asia, as shown by **Figure 2**. Of the 65 reactors now under construction around the world, 44 are in Asia (excluding the Middle East), while 14 are in Europe (including 10 in Russia), six in the Americas, and one in the Middle East (United Arab Emirates).[22] Planned or proposed nuclear power plants show a similar trend. Of the 483 planned or proposed reactors in **Figure 2**, nearly 60% (281) are in Asia, while 116 are in Europe, 44 in the Americas, and 36 in the Middle East. South Africa has proposed up to six new reactors.

The renewed worldwide interest in nuclear power has led to a possible expansion of the technology to currently non-nuclear nations. Ten of the countries that are currently building or formally planning reactor projects—Belarus, Egypt, Indonesia, Jordan, Kazakhstan, Poland, Thailand, Turkey, the United Arab Emirates, and Vietnam—have never operated nuclear power plants. Several other non-nuclear power countries have proposed building power reactors, including Bangladesh, Chile, Israel, Malaysia, North Korea, and Saudi Arabia.[23] Iran started full operation of its first nuclear power plant in 2012.

The International Atomic Energy Agency recently reported that the Fukushima Daiichi disaster had caused some developing countries to adopt a more cautious approach toward nuclear power, but the agency found that "interest continued among countries considering or planning for nuclear power introduction." Fourteen non-nuclear power countries in 2012 were "considering a nuclear programme to meet identified energy needs with a strong indication of intention to proceed," according to IAEA, unchanged from 2008 and 2010. Three had actually ordered new reactors, up from zero in 2008 and two in 2010, but six were in "active preparation for a possible nuclear power programme with no final decision," down from seven in 2008 and 2010.[24]

---

[19] World Nuclear Association, "World Nuclear Power Reactors & Uranium Requirements," September 2012, http://www.world-nuclear.org/info/reactors.html.

[20] International Energy Agency, *World Energy Outlook 2011*, p. 178, Annex A.

[21] World Nuclear Association Reactor Database, at http://www.world-nuclear.org/reference/reactorsdb_index.php.

[22] Iran's Bushehr Nuclear Power Plant was commissioned in September 2011, but it is not yet operating at full capacity. "Ceremony Marks Bushehr Commissioning," *World Nuclear News*, September 13, 2011.

[23] World Nuclear Association, http://www.world-nuclear.org/info/reactors.html.

[24] International Atomic Energy Agency, *International Status and Prospects for Nuclear Power 2012*, August 15, 2012, p. 9.

Nuclear Fuel Services Market

The possible upsurge in worldwide nuclear power plant construction has focused new attention on nuclear fuel production. Although the shutdown of nearly all Japanese nuclear power plants in the wake of the Fukushima disaster has weakened nuclear fuel demand, at least in the near term, chronic worldwide overcapacity in all phases of the nuclear fuel cycle appears to be ending, evidenced by higher prices for uranium and enrichment services in recent years.[25] The expected long-term trend toward tightening supplies has sparked plans for new fuel cycle facilities around the world and also renewed concerns about controls over the spread of nuclear fuel technology.

The nuclear fuel cycle begins with mining uranium ore and upgrading it to yellowcake. Because naturally occurring uranium lacks sufficient fissile ^{235}U to make fuel for commercial light-water reactors, the concentration of ^{235}U must be increased in a uranium enrichment plant several times above its natural level of 0.7%. A nuclear power plant operator or utility typically purchases yellowcake and contracts for its conversion to uranium hexafluoride, then enrichment, and finally fabrication into fuel assemblies (**Figure 1**). Commercial enrichment services are available in the United States, Europe, Russia, and Japan. Fuel fabrication services are even more widely available. While waiting for conversion, the yellowcake remains a fungible commodity that can be consigned by the reactor operator to any conversion plant and the product sent to any enrichment plant (within trade restrictions between countries).[26] The various stages of the nuclear fuel cycle are described below.

---

[25] Kapik, Michael, "Spot U3O8 Price Slips Below $51/lb," *NuclearFuel*, June 25, 2012, p. 1.
[26] IAEA, *Country Nuclear Fuel Cycle Profiles*, 2nd ed., 2005.

Figure 1. The Conceptual Nuclear Fuel Cycle

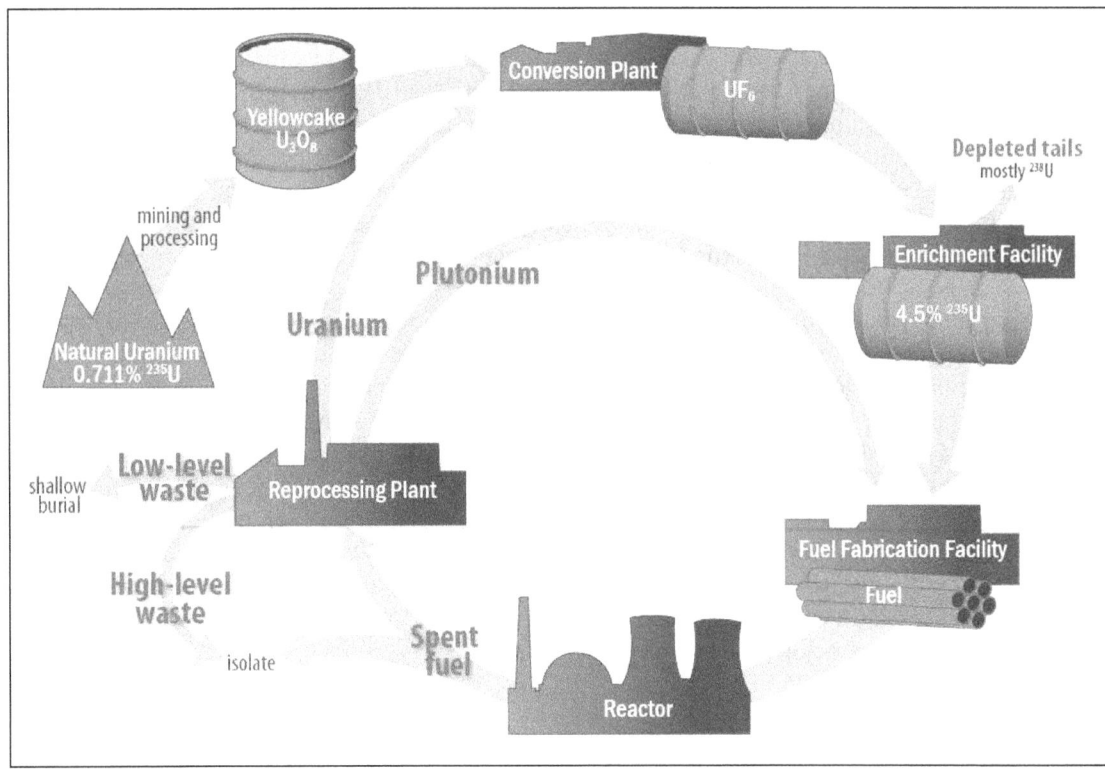

Yellowcake

Conventionally mined uranium ore (from open-pit and underground mines) is milled, then acid-leached to extract uranium oxide. The extract is then filtered, dried, and packaged as uranium yellowcake for shipment to a conversion plant. In-situ leaching avoids the mechanical mining steps by directly injecting solvents into the ore body through wells drilled from the surface. The dissolved uranium is pumped to the surface, where the uranium oxide is similarly processed into yellowcake for shipment.

The largest U.S. uranium reserves are located in Arizona, Colorado, New Mexico, Texas, Utah, and Wyoming, while several other states have smaller amounts.[27] According to the Energy Information Administration (EIA), 5 underground mines and 5 in-situ mines were operating in the United States in 2011, compared with a total of 8 the previous year and 18 in 2009. Total U.S. uranium production employment in 2011 was estimated at 1,191 person-years.[28] EIA reports 55 million pounds of U_3O_8 were purchased for U.S. nuclear power reactors in 2011, of which 9% was U.S. origin. The balance was made up in part by imported natural uranium and enriched uranium, including downblended highly enriched uranium (HEU), as discussed further below.[29]

---

[27] U.S. Energy Information Administration, *U.S. Uranium Reserves Estimates*, July 2010, http://www.eia.doe.gov/cneaf/nuclear/page/reserves/ures.html. Latest available data.

[28] U.S. Energy Information Administration, *2011 Domestic Uranium Production Report*, May 2012, http://www.eia.gov/uranium/production/annual/pdf/dupr.pdf.

[29] U.S. Energy Information Administration, *2011 Uranium Marketing Annual Report*, May 2012, http://www.eia.gov/uranium/marketing/pdf/2011umar.pdf.

A typical 1,000 MW light water reactor fuel load may require converting and enriching nearly 800,000 lb. of uranium "yellowcake" (U_3O_8). Approximately 64,400 metric tons (142 million lb.) of yellowcake was produced worldwide in 2011. That production was estimated to meet about 85% of worldwide demand for U_3O_8 for power generation.[30] Most of the difference between annual production and demand is covered by sales from former military uranium stockpiles.[31] The World Nuclear Association projects that demand for uranium could begin to exceed supply by about 2023, depending on the level of downblended weapons HEU, recycled uranium and plutonium, and other stockpiles, as well as production from new mines.[32]

Unlike gold or oil commodities, uranium yellowcake had not been offered through a formal market exchange until quite recently. Uranium price indicators had been developed by a small number of private business organizations, such as the World Nuclear Fuel Market (WNFM), TradeTech, and the Ux Consulting Company (UxC), that independently monitor uranium market activities, including offers, bids, and transactions. The price indicators are owned by and proprietary to the businesses that developed them.

NAC International (now a USEC Inc. subsidiary) established the WNFM to provide uranium price information in 1974. The WNFM membership comprises 93 companies representing 20 countries.[33] UxC and TradeTech maintain long- and short-term price indexes, which may be referenced by "market price" sales contracts; that is, sales contracts with pricing provisions that call for the future uranium delivery price to be equal to the market price at or around the time of delivery.[34]

In April 2007, the New York Mercantile Exchange (NYMEX) announced that it had partnered with the UxC to provide financially settled on- and off-exchange traded uranium futures contracts.[35] A NYMEX uranium futures contract's final settlement price is based on the UxC pricing index for yellowcake. Uranium futures contracts are available for trading on Chicago Mercantile Exchange Globex, and for clearing on NYMEX ClearPort.[36] The size of each contract is 250 lb., and prices are quoted in U.S. currency. The final settlement price is the spot month-end price published by UxC. A standard contract for trading physical uranium is currently under development.[37]

Uranium is typically mined outside the countries that use it. More than 70% of the world's production in 2011 came from Kazakhstan, Canada, Australia, and Niger (of which only Canada

---

[30] World Nuclear Association, *World Uranium Mining*, August 2012, http://www.world-nuclear.org/info/inf23 html.

[31] World Nuclear Association, *Military Warheads as a Source of Nuclear Fuel*, July 18, 2012, http://www.world-nuclear.org/info/inf13 html.

[32] Note: Metric tons is the unit of measurement for uranium fuel. One metric ton is approximately 2,200 pounds. World Nuclear Association, Uranium Markets, July 2010, http://www.world-nuclear.org/info/inf22 html.

[33] World Nuclear Fuel Market website, at http://www.wnfm.com. .

[34] See TradeTech website at http://www.uranium.info/index.php, and Ux Consulting Company website at http://www.uxc.com.

[35] CME Group, "UxC Uranium U308 Swap Futures," contract specifications, http://www.cmegroup.com/trading/metals/other/uranium_contract_specifications html.

[36] CME Globex is a global electronic trading platform for trading futures products. NYMEX ClearPort Clearing provides traders an interface where transactions are posted, margin requirements are calculated, and the transactions are processed by the clearinghouse.

[37] David Stellfox, "Industry Moves Closer to Having Standardized Uranium Contract," *NuclearFuel*, February 9, 2009.

uses nuclear power),[38] while more than half of the world's commercial reactors are in the United States, France, and Japan. But security of uranium supply, while always an underlying policy concern, has rarely been a real problem, because production vastly outstripped demand during the first three decades of the commercial nuclear power era—until about the mid-1980s.[39] As a result, a huge overhang of military and civilian stockpiles of uranium helped maintain a worldwide buyers' market.

Since the mid-1980s, however, world nuclear fuel requirements continued to rise while uranium exploration and production fell. By 2000, as U.S. spot-market prices hit bottom (at about $7 per pound), the western world's nuclear fuel requirements were twice the level of production. At that point, commercial stockpiles had been drawn down enough to begin putting pressure on U.S. spot prices, which rose slightly through 2003 and then dramatically (above $75 per pound) in 2007. Since then, prices have fluctuated and were about $50 per pound in late 2012.[40] The spot price represents about 20% of the market but provides an indicator of future contracts, which usually run three to seven years.[41]

Despite low worldwide exploration expenditures since the mid-1980s caused by oversupply and low prices, estimated uranium resources have trended upward over the long term. According to the OECD Nuclear Energy Agency (NEA), "new uranium resources can be readily identified" whenever prices rise due to tightening supply. "Regardless of the role that nuclear energy ultimately plays in meeting rising electricity demand, the uranium resource base described in this document is more than adequate to meet projected requirements. Meeting even high-case requirements to 2035 would consume less than half of the identified resources described in this volume," according to NEA.[42]

Conversion

In the conversion process, the yellowcake is purified, chemically reacted with hydrofluoric acid in various processes to form uranium hexafluoride (UF_6) gas, and then transferred into cylinders, where it cools and condenses to a solid. Uranium hexafluoride contains two isotopes of uranium—heavier ^{238}U and lighter, fissionable ^{235}U, which makes up ~0.7% of uranium by weight. The annual U.S. demand for yellowcake conversion is approximately 22,000 metric tons uranium (MTU). After conversion, the uranium hexafluoride is ready for enrichment.

Five major commercial conversion companies operate worldwide—in the United States, Canada, China, France, the United Kingdom, and Russia (**Table 2**). ConverDyn in Metropolis, IL, the only conversion plant operating in the United States, has an annual capacity of 15,000 MTU.[43]

---

[38] World Nuclear Association, *World Uranium Mining*, August 2012, http://www.world-nuclear.org/info/inf23.html.

[39] World Nuclear Association, *Uranium Markets*, July 2010, at http://www.world-nuclear.org/info/inf22.html.

[40] Michael Knapik, "Spot U3O8 Price May Now Be Below $46/lb," *NuclearFuel*, October 1, 2012, p. 1.

[41] World Nuclear Association, *Uranium Markets*, July 2010, at http://www.world-nuclear.org/info/inf22.html.

[42] Nuclear Energy Agency, Organization for Economic Cooperation and Development, *Uranium 2009: Resources, Production, and Demand*, pp. 13, 106.

[43] World Nuclear Association, "Uranium Enrichment," September 2012, http://www.world-nuclear.org/info/inf28.html.

Table 2. Commercial UF$_6$ Conversion Facilities

(metric tons uranium/year)

Country	Company	Facility	Capacity
Canada	Cameco	Port Hope	12,500
China	CNCC	Lanzhou	3,000
France	Comurhex	Peirrelatte	14,500
Russia	JSC	Irkutsk and Seversk	25,000
U.K.	Cameco	Springfields	6,000
U.S.	Converdyn	Metropolis	15,000
Total			76,000

Source: World Nuclear Association.

Enrichment

For use as fuel in light water reactors, uranium must be enriched in the isotope ^{235}U above its natural concentration of about 0.7%. By heating UF$_6$ to turn it into a gas, the enrichment process can take advantage of the slight difference in atomic mass between ^{235}U and ^{238}U. The typical enrichment process increases the ^{235}U concentration to 3%-5%, requiring about 10 lb. of uranium U$_3$O$_8$ to produce 1 lb. of enriched uranium hexafluoride (UF$_6$) product.

About 90% of the world's existing commercial reactors (all except heavy water reactors and some gas-cooled reactors) require enriched uranium fuel. Major enrichment plants are located in the United States, Russia, France, Great Britain, Germany, and the Netherlands, plus smaller plants in a few other countries (see **Table 2**). Thirty-one countries currently operate commercial nuclear power plants. Most countries, therefore, rely on enrichment services outside their borders. An enrichment plant to serve a country with only a few reactors would appear economically nonviable, given that a single large enrichment plant can supply up to 25% of the world market (currently estimated at 45,000 metric tons of separative work units, or SWUs).[44]

Commercial uranium enrichment employs either gaseous diffusion or high speed centrifuges. In gaseous diffusion, a thin, semiporous barrier holds back more of the heavier ^{238}U than the lighter ^{235}U. A series of cascading diffusers successively increases the ^{235}U concentration. Centrifuge enrichment spins the uranium hexafluoride gas at ultra-high speeds to separate the lighter ^{235}U. A series of cascading centrifuges successively enriches the gas in ^{235}U. Final enrichment will vary depending on the requirements of a specific reactor, normally up to about 5%.

Gaseous diffusion technology was first developed in the United States and later adopted by France and Britain. It is much more energy-intensive than the newer centrifuge enrichment process and has been largely phased out. The Georges Besse 1 gaseous diffusion plant in France shut down permanently in mid-2012. The only other major diffusion plant that is still operating, located in Paducah, KY, is scheduled to close in mid-2013.[45]

---

[44] Ruthane Neely and Jeff Combs, "Diffusion Fades Away," *Nuclear Engineering International*, September 2006, p. 24.

[45] USEC, Inc., "Five-Party Arrangement Extends Paducah Gaseous Diffusion Plant Enrichment Operations," press (continued...)

Uranium enrichment services are sold in kilograms (kg) or metric tons (1,000 kg) separative work units (SWU), which is a measure of the amount of work needed (in the thermodynamic sense) to enhance the ^{235}U concentration. The number of SWUs required to produce fuel depends on several factors: the quantity of fuel required, level of enrichment required, the initial enrichment of the feed (0.711% in the case of natural uranium), and the "tails assay," which is the ^{235}U concentration remaining in the depleted processing stream. For example, to produce 1 kg of uranium enriched to 3% ^{235}U, at a tails assay of 0.2% ^{235}U, 4.3 kg-SWU are used to process 5.5 kg of natural uranium.[46] The price of yellowcake is an important factor in enrichment demand. Under high price conditions, it may be economically preferable to expend more SWUs enriching a lesser quantity of yellowcake, thus leaving a lower tails assay.

Nuclear plant operators can buy uranium yellowcake and have it converted and enriched, or buy low-enriched uranium (LEU).[47] Commercial enrichment services are offered by a number of international sources (**Table 3**), with worldwide annual capacity of 57,350 metric tons SWU. In 2011, U.S. nuclear plant operators contracted in nine countries to provide 14,829 metric tons SWU, of which 16% was provided in the United States.[48]

Table 3. Operating Commercial Uranium Enrichment Facilities

(metric tons SWU/year in 2010)

Facility Name	Country	Process	Capacity
Tenex (multiple plants)	Russia	Centrifuge	23,000
Urenco (multiple plants)	Germany, Netherlands, UK, US	Centrifuge	13,000
Paducah	United States	Gaseous Diffusion	11,300
Eurodif (Georges Besse 1 and 2)	France	Gaseous Diffusion	8,500
CNNC (Hanzhun and Lanzhou)	China	Centrifuge	1,300
JNFL (Rokkasho)	Japan	Centrifuge	150
Other	Pakistan, Brazil, Iran	Centrifuge	100
Total			**57,350**

Source: World Nuclear Association.

The U.S. DOE and its predecessor agencies had operated gaseous diffusion enrichment plants in Oak Ridge, TN, Paducah, KY, and Portsmouth, OH, to produce high-enriched uranium used in the nuclear weapons program. The plants later produced low-enriched uranium for commercial nuclear power around the world, although production at the Oak Ridge K-25 enrichment site ceased in 1985. The Energy Policy Act of 1992 established the United States Enrichment

---

(...continued)

release, May 15, 2012, http://www.usec.com/news/five-party-arrangement-extends-paducah-gaseous-diffusion-plant-enrichment-operations.

[46] Thomas L. Neff, *The International Uranium Market*, Ballinger Publishing Co., 1984.

[47] LEU is generally defined as any enriched uranium product with less than 20% ^{235}U, while highly enriched uranium (HEU) contains 20% or more ^{235}U. Depleted uranium has ^{235}U levels below that of natural uranium.

[48] U.S. Energy Information Administration, Purchases of Enrichment Services by Owners and Operators of U.S. Civilian Nuclear Power Reactors by Origin Country and Year, 2007-2011, http://www.eia.gov/uranium/marketing/html/table16.cfmhttp://www.eia.doe.gov/cneaf/nuclear/umar/table16.html.

Corporation (USEC) as a government-owned corporation to take over DOE's uranium enrichment services business. The corporation was privatized as USEC Inc. in 1998. In 2001, USEC ceased uranium enrichment operations in Portsmouth and consolidated operations in Paducah.

In 2004, USEC announced plans to build the American Centrifuge Plant on the site of the Portsmouth gaseous diffusion plant. The new gas centrifuge enrichment plant is to have 11,500 centrifuges with an annual capacity of 3,800 metric tons SWU.[49] Construction of the plant was suspended in August 2009, with restart dependent on DOE approval of a loan guarantee.[50] DOE has been funding a demonstration program to prove the technical adequacy of the USEC centrifuge technology, which will be a prerequisite for the loan guarantee. The Continuing Appropriations Resolution for FY2013 (P.L. 112-175) included $100 million to continue funding the demonstration program.

Under the 1993 U.S.-Russian Federation Megatons to Megawatts program, highly enriched uranium from dismantled Russian nuclear warheads is converted into low-enriched uranium fuel for use in commercial U.S. nuclear power plants.[51] The HEU Agreement, as it is known, provides for the purchase through 2013 of 500 metric tons highly enriched uranium downblended to commercial grade low-enriched uranium (delivered as UF_6). USEC fulfills about half its enrichment supply contracts with Russian HEU that has been blended down into LEU, which totaled 858 metric tons in 2010.[52]

Urenco, a joint Dutch, German, and British enrichment consortium, was set up in the 1970s following the signing of the Treaty of Almelo. Urenco operates enrichment plants in Germany, the Netherlands, and the United Kingdom to supply customers in Europe, North America, and East Asia. Its U.S. affiliate, Louisiana Energy Services (LES), began startup operations at its newly constructed Urenco USA gas centrifuge enrichment plant in New Mexico on June 11, 2010.[53] The New Mexico facility is expected to produce 3,000 metric tons SWUs annually when it reaches full operational capacity in 2013—meeting approximately 25% of the current U.S. demand.[54] Urenco estimates that it provides around 25% of the world market share in enrichment services.[55]

Areva began starting up a gas centrifuge plant, Georges Besse II, on December 14, 2010, on the Tricastin nuclear site in France. Georges Besse II, which replaces the Georges Besse I gaseous diffusion plant at the same site, is to reach full capacity of 7,500 metric tons SWU by 2016.[56] Areva applied to NRC at the end of 2008 for a license to build and operate a similar gas

---

[49] *USEC to Site American Centrifuge Plant in Piketon, Ohio—Technology Expected To Be World's Most Efficient for Enriching Nuclear Fuel*, at http://www.usec.com/v2001_02/Content/News/NewsTemplate.asp?page=/v2001_02/Content/News/NewsFiles/01-12-04 htm.

[50] USEC Inc., "USEC Provides American Centrifuge Update," press release, November 2, 2009, http://www.usec.com/NewsRoom/NewsReleases/USECInc/2009/2009-11-02-USEC-Provides-American-Centrifuge htm.

[51] http://www.usec.com.

[52] USEC, "Megatons to Megawatts History," http://www.usec.com/megatonstomegawatts_history htm; USEC 2011 Annual Report, p. 12, http://bnymellon mobular.net/bnymellon/usu.

[53] LES, "Urenco USA, Now Operational, Receives Nuclear Regulatory Commission Feed Stock Authorisation," press release, June 11, 2010, http://www.urenco.com/content/325/URENCO-USA-now-operational-receives-Nuclear-Regulatory-Commission-feed-stock-authorisation-.aspx.

[54] Urenco website, http://www.urenco.com/fullArticle.aspx?m=1371.

[55] Urenco website, http://www.urenco.com/Content/2/About-URENCO.aspx.

[56] Areva, "Enrichment: Inauguration of the Georges Besse II Plant," press release, December 14, 2010, http://www.areva.com/EN/news-8653/enrichment-inauguration-of-the-georges-besse-ii-plant html.

centrifuge enrichment plant in Idaho, and NRC issued a final environmental impact statement for the facility in February 2011.[57]

A more advanced technology, laser enrichment, is being pursued by a consortium of GE and Hitachi called Global Laser Enrichment. NRC issued a license to construct and operate a commercial-scale laser enrichment plant in Wilmington, NC, on September 25, 2012, although the company has not decided whether to move forward with the project. It would have a capacity of up to 6,000 metric tons SWU.[58] However, concerns have been raised that laser enrichment technology could prove especially efficient in producing weapons-grade uranium and thus pose a proliferation risk.[59]

Fuel Fabrication

Like enrichment, fuel fabrication is a specialized service rather than a commodity transaction. Strict quality control is necessary at all stages of the fabrication process to prevent fuel failure and leakage during reactor operations. The first step after uranium enrichment is the conversion of low-enriched uranium hexafluoride (UF_6) to uranium dioxide (UO_2), which usually takes place at a fuel fabrication plant but may also be done at a separate conversion plant. At the fabrication plant, the UO_2 powder is then mixed for uniformity, blended with other ingredients as specified, compacted, granulated, and pressed into cylindrical pellets. The pellets are sintered (heated at high temperature) to a precise level of density. The pellets are ground to the correct size and loaded into zirconium alloy tubes about 12-15 feet long and half an inch in diameter to make nuclear fuel rods.[60]

The fuel rods are then attached together in arrays to form fuel assemblies. Most western-design reactors use assemblies with square arrays, with the number of rods in each assembly ranging from fewer than 100 to as many as 300. Fuel assembly arrays can also be circular, hexagonal, or triangular and also vary in numerous other design parameters as required by specific reactor designs. As described by one expert, "Nuclear fuel assemblies are highly engineered products, made especially to each customer's individual specifications. These are determined by the physical characteristics of the reactor, by the reactor operating and fuel cycle management strategy of the utility as well as national, or even regional, licensing requirements."[61]

Fuel fabrication services for light water reactors (LWRs) are offered by about a dozen suppliers operating in 14 countries at about 20 facilities. In 2011, the World Nuclear Association estimated that annual worldwide LWR fuel fabrication capacity stood at about 13,000 metric tons (of total

---

[57] Nuclear Regulatory Commission, "AREVA Enrichment Services, LLC Gas Centrifuge Facility," http://www.nrc.gov/materials/fuel-cycle-fac/arevanc.html.

[58] GE Hitachi, "GE Hitachi Nuclear Energy's Global Laser Enrichment Receives Nuclear Regulatory Commission License for Uranium Plant," press release, September 25, 2012, http://www.genewscenter.com/News/GE-Hitachi-Nuclear-Energy-s-Global-Laser-Enrichment-Receives-Nuclear-Regulatory-Commission-License-for-Uranium-Plant-3b4a.aspx.

[59] R. Scott Kemp, "SILEX and Proliferation," *Bulletin of the Atomic Scientists*, web edition, July 30, 2012, http://thebulletin.org/web-edition/features/silex-and-proliferation.

[60] "Nuclear Fuel Fabrication," in *Nuclear Engineering Handbook*, ed. Kenneth D. Kok (Boca Raton, FL: CRC Press, 2009), pp. 279-291.

[61] Steve Kidd, Director of Strategy and Research, *World Nuclear Association*, "Fuel Fabrication – Outside of the Fuel Cycle?," Nuclear Engineering International, January 14, 2010, http://www.neimagazine.com/story.asp?storyCode=2055169.

uranium content), exceeding demand of 7,000 tons by 85%. The oversupply has existed for many years, although demand is projected to rise to 9,500 tons by 2020. Three LWR fuel fabrication facilities are located in the United States: Areva Inc. in Richland, WA; Global Nuclear Fuel in Wilmington, NC; and Westinghouse Electric in Columbia, SC. Several facilities also provide fuel for other types of reactors.[62]

Final Stages of the Fuel Cycle

The final stages of the nuclear fuel cycle take place after nuclear fuel assemblies have been loaded into a reactor. In the reactor, the uranium 235 (^{235}U) splits, or fissions, releasing energy and neutrons and creating fission products (highly radioactive fragments of ^{235}U nuclei). The neutrons may cause other ^{235}U nuclei to fission, creating a nuclear chain reaction. Some neutrons are also absorbed by ^{238}U nuclei to eventually create plutonium 239 (^{239}Pu), which itself may then fission.

After several years in the reactor, fuel assemblies will build up too many neutron-absorbing fission products and become too depleted in fissile ^{235}U to efficiently sustain a nuclear chain reaction. At that point, the assemblies are considered spent nuclear fuel and removed from the reactor. Spent fuel typically contains about 1% ^{235}U, 1% plutonium, 4% fission products and other radioactive waste, and the remainder ^{238}U.

What to do with spent fuel has proved highly contentious. One option is direct disposal in a deep geologic repository to isolate spent fuel for the hundreds of thousands of years that it may remain hazardous. Another option is to reprocess the spent fuel to separate the uranium and plutonium for use in new fuel. Supporters of reprocessing, or recycling, contend that it could greatly reduce the volume and longevity of nuclear waste while vastly expanding the amount of energy extracted from the world's uranium resources. Opponents contend that commercial use of separated plutonium—a key material in nuclear weapons as well as reactor fuel—increases the worldwide risk of nuclear weapons proliferation.

Commercial-scale spent fuel reprocessing is currently conducted in France, Britain, and Russia. The ^{239}Pu they produce is blended with uranium to make mixed-oxide (MOX) fuel, in which the ^{239}Pu largely substitutes for ^{235}U. Two French reprocessing plants at La Hague can each reprocess up to 850 metric tons of spent fuel per year, while Britain's THORP facility at Sellafield has a capacity of 900 metric tons per year. Russia has a 400-ton plant at Ozersk, and Japan is building an 800-ton plant at Rokkasho to succeed a 90-ton demonstration facility at Tokai Mura. Britain and France also have older plants to reprocess gas-cooled reactor fuel, and India has four plants with a total annual capacity of 330 tons.[63] About 200 metric tons of MOX fuel is used annually, about 2% of new nuclear fuel,[64] equivalent to about 2,000 metric tons of mined uranium.[65]

However, the benefits of reprocessing spent fuel to make MOX fuel for today's nuclear power plants are modest. Existing commercial light water reactors use ordinary water to slow down, or

---

[62] World Nuclear Association, "Nuclear Fuel Fabrication," September 2011, http://www.world-nuclear.org/info/nuclear_fuel_fabrication-inf127.html.

[63] World Nuclear Association, *Processing of Used NuclearFuel*, May 2012, at http://www.world-nuclear.org/info/inf69.html.

[64] World Nuclear Association, *Mixed Oxide Fuel (MOX)*, May 2012, at http://www.world-nuclear.org/info/inf29.html.

[65] World Nuclear Association, *Uranium Markets*, July 2010.

"moderate," the neutrons released by the fission process. The relatively slow (thermal) neutrons are highly efficient in causing fission in certain isotopes of heavy elements, such as ^{235}U and ^{239}Pu.[66] Therefore, relatively low quantities of those isotopes are needed in nuclear fuel to sustain a nuclear chain reaction, allowing the use of low-enriched uranium.[67] The downside is that thermal neutrons cannot efficiently induce fission in more than a few specific isotopes. In today's commercial reactors, therefore, the buildup of non-fissile plutonium and other isotopes sharply limits the number of reprocessing cycles before the recycled fuel can no longer sustain a nuclear chain reaction and must be stored or disposed of.

In contrast, "fast" neutrons, which have not been moderated, are less effective in inducing fission than thermal neutrons but can induce fission in all actinides, including all plutonium isotopes. Therefore, nuclear fuel for a fast reactor must have a higher proportion of fissionable isotopes than a thermal reactor to sustain a chain reaction, but a larger number of different isotopes can constitute that fissionable proportion.

A fast reactor's ability to fission all actinides (actinium and heavier elements), makes it theoretically possible to repeatedly separate those materials from spent fuel and feed them back into the reactor until they are entirely fissioned. Fast reactors are also ideal for "breeding" the maximum amount of ^{239}Pu from ^{238}U, eventually converting virtually all of natural uranium to useable nuclear fuel.[68] Current reprocessing programs are generally viewed by their proponents as interim steps toward a commercial nuclear fuel cycle based on fast reactors. However, critics point out that fast reactor technology has proven difficult to commercialize.[69]

Waste Disposal and Energy Security

Reprocessing of spent fuel from fast breeder reactors has long been the ultimate goal of nuclear power supporters. As noted above, fast reactors (operated either as breeders or non-breeders) can largely eliminate plutonium from nuclear waste and greatly extend uranium supplies. But opponents contend that such potential benefits are not worth the costs and nonproliferation risks.

Removing uranium from spent nuclear fuel through reprocessing would eliminate most of the volume of radioactive material requiring disposal in a deep geologic repository. In addition, the removal of plutonium and conversion to shorter-lived fission products would eliminate most of the long-term (post-1,000 years) radioactivity in nuclear waste. But the waste resulting from reprocessing would have nearly the same short-term radioactivity and heat as the original spent fuel, because the reprocessing waste consists primarily of fission products, which generate most of the near-term radioactivity and heat in spent fuel. Because heat is the main limiting factor on

---

[66] Isotopes are atoms of the same chemical element but with different numbers of neutrons in their nuclei.

[67] Reactors moderated with heavy water, which absorbs fewer neutrons than ordinary, "light" water, can operate with natural, unenriched uranium.

[68] The core of a breeder reactor is configured so that more fissile ^{239}Pu is produced from ^{238}U than the amount of fissile material initially loaded into the core that is consumed (^{235}U or ^{239}Pu). In a breeder, therefore, enough fissile material could be recovered through reprocessing to refuel the reactor and to provide fuel for additional breeders. The core of a fast reactor can also be configured to produce less ^{239}Pu than fissile material consumed, if the primary goal is to eliminate ^{239}Pu from spent fuel.

[69] Thomas B. Cochran, et al., Fast Breeder Reactor Programs: History and Status, International Panel on Fissile Materials Research Report 8, February 2010, http://www.ipfmlibrary.org/rr08.pdf.

repository capacity, conventional reprocessing would not provide major disposal benefits in the near term.

To address that problem, various proposals have been made to further separate the primary heat-generating fission products—cesium 137 and strontium 90—from high-level waste for separate storage and decay over several hundred years. Such a process could greatly increase repository capacity, although it would require an alternative secure storage system for the cesium and strontium for hundreds of years.

Energy security has been a primary driving force behind the development of nuclear energy, particularly in countries such as France and Japan that have few natural energy resources. Recent cutoffs in oil and gas around the world have underscored the instability of oil and gas supply, which could be mitigated by nuclear energy. For example, in 2006, a natural gas price dispute between Russia and Ukraine resulted in a temporary cutoff of natural gas to Western and Central Europe; in 2007, price disputes between Russia and Azerbaijan and Belarus caused a temporary cutoff in oil to Russia from Azerbaijan and in oil from Russia to Germany, Poland, and Slovakia. Moreover, temporary production shutdowns in the Gulf of Mexico and the Trans-Alaskan pipeline, instability in Nigeria, and nationalization of oil and gas fields in Bolivia in 2006 all raised concerns about oil and gas supplies and worldwide price volatility. Relative to gas and oil, the ability to stockpile uranium is widely seen as offering greater assurances of weathering potential cutoffs. However, nuclear electricity in most cases is not directly substitutable for oil's most common use, as transportation fuel.

Worldwide uranium resources are generally considered to be sufficient for at least several decades. Uranium supply is highly diversified, with uranium mining spread across the globe, while uranium conversion, enrichment, and fuel fabrication are more concentrated in a handful of countries. But because most reactors around the world rely at least partly on foreign sources of uranium and nuclear fuel services, nuclear reactors nearly everywhere face some level of supply vulnerability. To mitigate such concern, countries such as China, India, and Japan are seeking to secure long-term uranium contracts to support nuclear expansion goals. Efforts are underway to establish an international nuclear fuel bank to provide greater certainty in fuel supplies, as discussed in the next section.

Ultimately, only the development of breeder reactors, reprocessing, and plutonium fuel fabrication could provide complete nuclear energy independence for most countries. This remains the long-term goal of resource-poor France and Japan (which is reevaluating its program in the wake of Fukushima), and Russia as well, although their research and development programs have faced numerous obstacles and schedule slowdowns.

Proposals on the Fuel Cycle

Proposals addressing access to the full nuclear fuel cycle have ranged from seeking a formal commitment to forswear enrichment and reprocessing technology, to a de facto approach in which a state does not operate fuel cycle facilities but makes no explicit commitment to give them up, to no restrictions at all. Current proposals generally aim to persuade countries not to develop their own fuel production capabilities by providing economically attractive alternatives that allay concerns about politically motivated interruption to fuel supply. Most proposals focus on this front-end problem, dealing with fuel supply and production issues. Ultimately, these proposals are aimed at preventing an increase in the number of states that would be capable of producing weapons-usable nuclear material.

The main proposals under discussion are outlined here in categories—those addressing the full fuel cycle, those addressing the "front-end" or assurance of fuel supply issues (including fuel banks and multilateral enrichment services), and those focusing on the "back-end" or waste disposal solutions.

Comprehensive Proposals

El Baradei Proposal (2003)

IAEA Director General Mohamed El Baradei in 2003 proposed a three-pronged approach to limiting the processing of weapon-usable material (separated plutonium and high-enriched uranium) in civilian nuclear fuel cycles.[70] First, he would place all enrichment and reprocessing facilities under multinational control. Second, he would develop new nuclear technologies that would not produce weapons-usable fissile material—in other words, "the holy grail" of a proliferation-resistant fuel cycle. In his October 2003 article in the *Economist* where he laid out these ideas, El Baradei maintained, "This is not a futuristic dream; much of the technology for proliferation-resistant nuclear-energy systems has already been developed or is actively being researched." Third, El Baradei proposed considering "multinational approaches to the management and disposal of spent fuel and radioactive waste." El Baradei did not place any nonproliferation requirements on participation, but instead suggested that the system "should be inclusive; nuclear-weapon states, non-nuclear-weapon states, and those outside the current non-proliferation regime should all have a seat at the table." Further, he noted that a future system should achieve full parity among all states under a new security structure that does not depend on nuclear weapons or nuclear deterrence.

IAEA Experts Group/INFCIRC/640

In February 2005, an Expert Group commissioned by IAEA Director General El Baradei to explore these ideas presented a report, "Multilateral Approaches to the Nuclear Fuel Cycle."[71] The Expert Group studied several possible approaches to securing the operation of proliferation-sensitive nuclear fuel cycle activities (uranium enrichment, reprocessing and spent fuel disposal, and storage of spent fuel) and analyzed the incentives and disincentives for states to participate. The report reviewed relevant past and present experience. The Group's suggested approaches included the following:

- Reinforce existing market mechanisms by providing additional supply guarantees by suppliers and/or the IAEA (fuel bank).

- Convert existing facilities to multinational facilities.

- Create co-managed, jointly owned facilities.

---

[70] "Towards a Safer World," at http://f40.iaea.org/worldatom/Press/Statements/2003/ebTE20031016.shtml.

[71] "Multilateral Approaches to the Nuclear Fuel Cycle: Expert Group Report submitted to the Director General of the International Atomic Energy Agency," February 22, 2005, (INFCIRC/640). Available at http://www.iaea.org/Publications/ocuments/Infcircs/2005/infcirc640.pdf.

The Group concluded that "in reality, countries will enter into multilateral arrangements according to the economic and political incentives and disincentives offered by these arrangements."[72] The report noted that no legal framework existed for requiring states to join supply assurance arrangements.

In September 2006, the IAEA sponsored a conference entitled "New Framework for the Utilization of Nuclear Energy in the 21st Century: Assurances of Supply and Non-Proliferation," which addressed proposals to provide fuel assurances. The IAEA presented a report on fuel assurance options at the June 2007 Board of Governors meeting analyzing the various proposals put forth to date.[73] A potential framework for nuclear supply assurances could have three stages: (1) existing market arrangements; (2) back-up commitments by suppliers in case of a politically motivated interruption of supply if nonproliferation criteria are met; (3) a physical LEU material reserve.[74] The report emphasizes that participation in these arrangements should be voluntary, that progress on this question will be incremental and that many options should be explored to give consumer states sufficient choices to meet their needs.

President Bush's 2004 Proposal

In a speech at the National Defense University on February 11, 2004, President Bush said the world needed to "close a loophole" in the NPT that allows states to legally acquire the technology to produce nuclear material which could be used for a clandestine weapons program. To remedy this, he proposed that the 40 members of the Nuclear Suppliers Group (NSG) should "refuse to sell enrichment and reprocessing equipment and technologies to any state that does not already possess full-scale, functioning enrichment and reprocessing plants."[75] President Bush also called on the world's leading nuclear fuel services exporters to "ensure that states have reliable access at reasonable cost to fuel for civilian reactors, so long as those states renounce enrichment and reprocessing." President Bush's 2004 proposal was the only one that calls for countries to explicitly "renounce" pursuit of enrichment or reprocessing technologies in exchange for reliable access to nuclear fuel, and proved controversial. Many non-nuclear-weapon states saw this as an attempt to limit their access to the peaceful use of nuclear energy provided for under Article IV of the NPT.

Russia's "Global Nuclear Power Infrastructure"

In January 2006, Russian President Vladimir Putin proposed the Global Nuclear Power Infrastructure initiative that would include four kinds of cooperation: creation of international uranium-enrichment centers (IUECs), international centers for reprocessing and storing spent nuclear fuel, international centers for training and certifying nuclear power plant staff, and an international research effort on proliferation-resistant nuclear energy technology. The international fuel cycle centers would be under joint ownership and co-management. They would

---

[72] Ibid., p. 98.

[73] International Atomic Energy Agency, *Possible New Framework for the Utilization of Nuclear Energy: Options for Assurance of Supply of Nuclear Fuel*, June 2007.

[74] Tariq Rauf, "Realizing Nuclear Fuel Assurances: Third Time's the Charm," Presentation to the Carnegie International Nonproliferation Conference, June 24, 2007, at http://www.carnegieendowment.org/files/fuel_assurances_rauf.pdf.

[75] "President Announces New Measures to Counter the Threat of WMD," February 11, 2004, at http://www.whitehouse.gov/news/releases/2004/02/20040211-4.html.

be commercial joint ventures (that is, no state financing), with advisory boards consisting of government, industry, and IAEA professionals. The IAEA would not have a vote on these boards, but would play an advisory role, while also certifying the fuel provision commitments.

Recipient countries under Putin's proposal would receive fuel cycle services, but access to sensitive technology would stay in the hands of the supplier state. Russia has offered a similar arrangement to Iran—to jointly enrich uranium on Russian territory. Iran has not yet accepted this offer, but it is still part of ongoing negotiations with Iran over its nuclear program. Russia made the return of spent fuel from the Bushehr nuclear plant in Iran a condition of supply, so that no plutonium can be extracted from the spent fuel.[76] To date, progress has been made is establishing the IUEC and in establishing an LEU reserve at Angarsk (see below). Russia has established a center of excellence for nuclear personnel at Obninsk, and participates in joint research efforts on fast reactors such as the International Project on Innovative Nuclear Reactors and Fuel Cycles (INPRO).

Assurance of Fuel Supply: Supplier Guarantees

The following proposals focus on back-up fuel supply assurances designed to complement, but not impact or supplant, the commercial uranium market.

Six Country Concept

In May 2006, six governments—France, Germany, the Netherlands, Russia, the United Kingdom, and the United States—proposed a "Concept for a Multilateral Mechanism for Reliable Access to Nuclear Fuel"[77] (referred to here as the Six Country Concept). This proposal reportedly developed from a U.S. initiative following President Bush's 2004 proposal. It would not require states to forgo enrichment and reprocessing, but participation would be limited to those states that did not currently have enrichment and reprocessing capabilities.

The Six Country Concept calls for a multi-tiered backup mechanism to ensure the supply of low enriched uranium (LEU) for nuclear fuel. The proposal would work as follows: (1) A commercial supply relationship is interrupted for reasons other than nonproliferation; (2) The recipient or supplier state can approach the IAEA to request backup supply; (3) The IAEA would rule out commercial or technical reasons for interruption (to avoid a market disruption) and assess whether the recipient meets the following qualifications: it must have a comprehensive safeguards agreement and Additional Protocol in force; it must adhere to international nuclear safety and physical protection standards; and it is not pursuing sensitive fuel cycle activities (which are not defined); (4) The IAEA would facilitate new arrangements with alternative suppliers.

Two mechanisms were proposed to create multiple tiers of assurances: including a standard backup provision in commercial contracts, and establishing reserves of LEU (not necessarily held by the IAEA, but possibly with rights regarding the use of the reserves). The Six Country

---

[76] "The Last Word: Sergei Kirienko," *Newsweek*, February 20, 2006 issue, at http://www msnbc msn.com/id/11299203/site/newsweek/.

[77] "Concept for a Multilateral Mechanism for Reliable Access to Nuclear Fuel," Proposal as sent to the IAEA from France, Germany, the Netherlands, Russia, Ireland, and the United States, May 31, 2006, IAEA GOV/INF/2006/10. Available at http://www-pub.iaea.org/MTCD/Meetings/PDFplus/2006/cn147_ConceptRA_NF.pdf.

Concept addressed several future options, all of which are longer-term in nature. They include providing reliable access to existing reprocessing capabilities for spent fuel management; multilateral cooperation in fresh fuel fabrication and spent fuel management; international enrichment centers; and new fuel cycle technology development that could incorporate fuel supply assurances.

World Nuclear Association

In May 2006, the private-sector World Nuclear Association (WNA) Working Group on Security of the International Nuclear Fuel Cycle outlined proposals for assuring front-end and back-end nuclear fuel supplies.[78] Like the Six Country Concept, the WNA proposal envisions a system of supply assurances that starts first with normal market procedures attempting to reestablish nuclear fuel supply after interruptions. Also similar to the Six Country proposal, a pre-established network of suppliers could be triggered through the IAEA if supply were interrupted for political reasons. If that network then failed, stocks held by national governments could be used.

The first tier of assurances, therefore, is through commercial suppliers. The second level of supply commitment would use a "standard backup supply clause" in enrichment contracts, supported by governments and the IAEA. "To ensure that no single enricher is unfairly burdened with the responsibility of providing backup supply, the other (remaining) enrichers would then supply the contracted enrichment in equal shares under terms agreed between the IAEA and the enrichers," according to the proposal.

For fuel fabrication, a backup supply system would be more complicated, according to the WNA report. "Because fuel design is specific to each reactor design, an effective mechanism would require stockpiling of different fuel types/designs. The cost of such a mechanism could thus be substantial," according to the report. However, WNA noted that unlike uranium enrichment technology, uranium fuel fabrication is not of proliferation concern.

The WNA report also noted the need for back-end nuclear fuel cycle supply assurances, to prevent a future scenario in which reprocessing technologies spread as nuclear power programs expand. The report recommends that a clear option to reprocess spent fuel at affordable prices be offered to states that do not have indigenous reprocessing programs. Such assurances would be part of a longer-term approach.

Japan's IAEA Standby Arrangements System

Japan presented a "complementary proposal" to the Six Country Concept at the IAEA in September 2006.[79] Japan's concerns with the Six Country Concept centered on the implication that it would deny the right for states to use nuclear technology for commercial purposes and because it assured the supply only of LEU, rather than all front-end nuclear fuel cycle services. Japan proposed instead to create an "IAEA Standby Arrangements System" that would act as an early warning system to prevent a break in supply to recipients. With a list of supply capacities from each state updated annually and a virtual bank of front-end fuel cycle services (from natural uranium to fuel fabrication), the IAEA would facilitate supply to recipient states before supply

---

[78] WNA's report is available at http://www.world-nuclear.org/reference/pdf/security.pdf.

[79] Full text of proposal (INFCIRC/683) at http://www.iaea.org/Publications/Documents/Infcircs/2006/infcirc683.pdf.

was completely stopped. States determined by the IAEA Board of Governors to be in good non-proliferation standing by the IAEA could participate.

UK "Nuclear Fuel Assurance (NFA)"

The United Kingdom has proposed a political assurance of non-interference in the delivery of commercial nuclear contracts, called a nuclear fuel assurance (NFA). This concept incorporates an earlier proposal that would creates "enrichment bonds" to give advance assurance of export approvals for nuclear fuel to recipient states. As the UK Prime Minister's report, *Road to 2010*, summarized the NFA: "The UK's Nuclear Fuel Assurance is complementary to other proposals put forward and provides a guarantee that export licences for nuclear fuel enrichment services would only be withheld in the event of non-compliance with non-proliferation obligations."[80]

The NFA would be a government-to-government agreement between supplier state or states and the recipient state, with the IAEA as co-signatory. The supplier government would guarantee that, subject to the IAEA's determination that the recipient was in good nonproliferation standing, national enrichment providers will be given the necessary export approvals to supply the recipient states. It is a transparent legal mechanism designed to give further credible assurance of supply with a "prior consent to export" arrangement. The IAEA would make the final decision on whether conditions had been met to allow the export of LEU.[81] A model agreement to be used as a standard for an NFA could be adopted by the Board of Governors under the proposal.

Assurance of Fuel Supply: Fuel Reserves

A fuel reserve is meant to appeal to countries concerned about a possible cut-off in their nuclear fuel supply for reasons unrelated to nonproliferation, such as a non-commercial or political dispute with the supplier country. Two fuel banks have been approved by the IAEA Board of Governors. The IAEA Board approved terms for a Russian-operated "fuel reserve" in November 2009. The Board approved terms for an IAEA owned and managed fuel bank in December 2010. Additionally, the United States has declared that it would downblend excess military HEU to LEU and hold it in reserve as part of the Six Country Concept, under the Department of Energy's American Assured Fuel Supply program. These three proposals are discussed below.

U.S. LEU Fuel Reserve, the "American Assured Fuel Supply" (AFS)

In 2005, then-Secretary of Energy Samuel Bodman announced that 17.4 metric tons of U.S. surplus highly enriched uranium would be downblended to low-enriched uranium to be used as a U.S. fuel reserve. The goal of the U.S. reserve is to supply fuel in the event of a disruption unrelated to proliferation. Secretary Bodman described the U.S. reserve as supporting the "twin goals of expanding the use of nuclear power and curbing nuclear proliferation," and said its purpose was to "help countries to pursue nuclear power confidently, without the burden of producing their own fuel, while curbing the spread of sensitive technology."[82]

---

[80] *The Road to 2010: addressing the nuclear question in the 21st century*, July 2009, http://www.cabinetoffice.gov.uk/reports/roadto2010.aspx.

[81] INFCIRC/707, June 4, 2007.

[82] "DOE/NNSA Reliable Fuel Supply Gains Momentum," NNSA Press Release, November 7, 2006.

The material designated for the U.S. reserve would be kept under national control, and not be part of the IAEA fuel bank.[83] DOE's *Federal Register* notice published in August 2011 noted that the AFS is intended to complement the international nuclear fuel bank. The reserve will be available to both domestic and foreign nuclear power plants "after all other market options are exhausted."[84] Recipients must "meet certain nonproliferation criteria." Stringent U.S. requirements on U.S.-origin material, pursuant to the 1954 Atomic Energy Act (as amended), include safeguards in perpetuity, prior consent for enrichment and reprocessing, and the right of return should a non-nuclear-weapon state detonate a nuclear explosive device.[85] Foreign recipients facing a supply disruption would obtain AFS fuel through their U.S. supplier with appropriate licenses. The *Federal Register* notice further specifies that the price of the LEU would be established "at the time of delivery using commercially acceptable market indices."[86]

DOE's Fissile Material Disposition program manages this initiative, formerly called "Reliable Fuel Supply." WesDyne International, LLC, and Nuclear Fuel Services, Inc., were awarded a contract in 2007 to downblend and store the material. Nuclear Fuel Services began downblending in December 2009 at its facility in Erwin, TN and is expected to complete the work by the end of 2012. The 17.4 MT of HEU will produce about 290 MT of low enriched uranium. According to an NNSA press release, WesDyne will sell a "small fraction" of the resulting low enriched uranium on the market over a three- to four-year period to cover the project's costs.[87] Of the 290 MT, approximately 230 MT will make up the reserve, enough for approximately six reactor core reloads for an average 1,000 MW reactor.[88]

In addition to the downblending of 17.4 MT of HEU, an additional 12.1 MT is to be downblended to approximately 220 MT of LEU "to provide assurance of fuel supply to utilities participating in the MOX program for the disposition of surplus weapons plutonium."[89] This tranche is expected to be downblended by the end of 2012.

LEU from both of the above HEU disposition programs will be stored until needed at Westinghouse's Columbia Fuel Fabrication Facility in Columbia, SC.

IAEA LEU Fuel Bank

The IAEA Board of Governors approved an IAEA-owned and managed LEU fuel bank on December 3, 2010. The reserve would consist of "enough LEU to meet the fuel fabrication needs of one full core of a 1,000 MW(e) pressurized water reactor, or three annual reloads of fuel."[90] In

---

[83] "News Analysis: The Growing Nuclear Fuel-Cycle Debate," Arms Control Today, November 2006. Available at http://www.armscontrol.org/act/2006_11/NAFuel.asp.

[84] "Notice of Availability: American Assured Fuel Supply," *Federal Register*, Vol. 76, No. 160, August 18, 2011.

[85] See also CRS Report RS22937, *Nuclear Cooperation with Other Countries: A Primer*, by Paul K. Kerr and Mary Beth Nikitin.

[86] The Notice of Availability further details the process for requesting an LEU purchase from the AFS. "Notice of Availability: American Assured Fuel Supply," *Federal Register*, Vol. 76, No. 160, August 28, 2011.

[87] "NNSA Awards Contract for Reliable Fuel Supply Program," NNSA Press Release, June 29, 2007.

[88] "Notice of Availability: American Assured Fuel Supply," Federal Register, Vol. 76, No. 160, August 28, 2011.

[89] "NNSA Announces Contract to Downblend 12 Metric Tons of Surplus Highly Enriched Uranium," NNSA Press Release, June 23, 2009.

[90] "Factsheet: IAEA Low Enriched Uranium Reserve," International Atomic Energy Agency, http://www.iaea.org/Publications/Factsheets/English/iaea_leureserve.html.

order to access LEU from the bank, the Director General must determine that the country has met the following conditions: the state experiencing interruption in supply is unable to acquire the fuel through the commercial market or other means; there are no outstanding safeguards implementation or diversion issues in the requesting state; and a comprehensive safeguards agreement is in place in the requesting country. Additional nonproliferation conditions are placed on the fuel once it is transferred. The recipient country is to pay the IAEA at the current market rate prior to transfer of the LEU. The government of Kazakhstan informed the IAEA in May 2009 (INCIRC/753) and in January 2010 (INFCIRC/782) that it would be willing to host the fuel bank on its territory.[91] The location for the reserve has not yet been finalized, and an agreement between the host government and the IAEA would need to be concluded.

The IAEA fuel bank began with monetary pledges by donors. In September 2006, former Senator Sam Nunn, co-chairman of the Nuclear Threat Initiative (NTI),[92] announced NTI's pledge of $50 million as seed money to create a low-enriched uranium stockpile owned and managed by the IAEA. NTI believes that the establishment of such an LEU reserve would assure an international supply of nuclear fuel on a non-discriminatory, non-political basis to recipient states. Provision of the NTI money was contingent on the IAEA taking the necessary preparatory actions to establish the reserve and on contribution of an additional $100 million or an equivalent value of LEU by one or more IAEA Member States.[93] The latter condition was met in March 2009. The U.S. Congress approved $50 million for an international fuel bank in December 2007 (see below). Norway pledged $5 million to the fuel bank in February 2008.[94] The United Arab Emirates announced a contribution of $10 million on August 1, 2008.[95] The European Union pledged 25 million euros in December 2008,[96] and Kuwait pledged $10 million in March 2009.[97] No policy conditions were set by donor countries—policy questions were meant to be solved by the IAEA and member states. Kazakhstan proposed in May 2009 that it host the fuel bank.

The IAEA secretariat drew up draft plans for the fuel bank, which were first presented to the Board of Governors at its June 2009 meeting along with a proposal for a Russian-hosted bank and the German-proposed multilateral enrichment project. At that time, some developing countries reportedly rejected the director general's proposal to negotiate details and approve these arrangements at the September 2009 Board meeting. Opponents and skeptics cited concerns that

---

[91] "Communication dated 11 January 2010 received from the Permanent Mission of the Republic of Kazakhstan to the Agency enclosing a positions regarding the establishment of IAEA nuclear fuel banks" (INFCIRC/782), January 15, 2010.

[92] Nuclear Threat Initiative is a private organization founded in 2001 by Mr. Ted Turner and former Senator Sam Nunn. It is now classified as a 501(c)3 public charity.

[93] Nuclear Threat Initiative Commits $50 million to Create IAEA Nuclear Fuel Bank, International Atomic Energy Agency Press Release, September 19, 2006. Available at http://www nti.org/c_press/release_IAEA_Fuelbank_091906.pdf.

[94] "Norway Contributes $5 Million to IAEA Nuclear Fuel Reserve," Norwegian Ministry of Foreign Affairs Press Release No. 027/08, February 27, 2008. http://www.regjeringen no/en/dep/ud/Press-Contacts/News/2008/fuelreserve.html?id=50210.

[95] "UAE Commits $10 Million to Nuclear Fuel Reserve Proposal," IAEA Press Release, August 7, 2008 http://www.iaea.org/NewsCenter/News/2008/uaecontribution.html; "UAE Commitment Gives NTI/IAEA Fuel Bank Critical Momentum," NTI Press Release, August 7, 2008. http://www nti.org/c_press/release_UAE%20fuel%20bank%2080708.pdf.

[96] "Sam Nunn Applauds EU Contribution to IAEA Fuel Bank," NTI Press Release, December 10, 2008. http://www.nti.org/c_press/statement_Nunn_EU_fuel_bank_121008.pdf.

[97] "Multinational Fuel Bank Reaches Key Milestone," IAEA Staff Report, March 6, 2009. http://www.iaea.org/NewsCenter/News/2009/fbankmilestone.html.

their legal rights under the NPT to develop fuel cycle facilities would be infringed upon if such facilities were established. Proponents counter that these arrangements would be optional, and are meant to give countries alternatives to developing their own fuel cycle capabilities.[98]

Congressional Approval

The National Defense Authorization Act for Fiscal Year 2008 (P.L. 110-181) authorized $50 million to be appropriated to the Department of Energy for the "International Atomic Energy Agency Nuclear Fuel Bank."[99] The report supports the establishment of a fuel bank and notes that "additional work will be required in order to provide appropriate guidance to the executive branch regarding criteria for access by foreign countries to any fuel bank established at the IAEA with materials or funds provided by the United States."[100]

Both the House (H.R. 2641) and Senate (S. 1751) Energy and Water Appropriations bills for FY2008 recommended funding for an international nuclear fuel bank under the IAEA, and proposed making available $100 million and $50 million respectively. The Consolidated Appropriations Act for FY2008, which became P.L. 110-161 on December 26, 2007, provided that $50 million should be available until expended for "the contribution of the United States to create a low-enriched uranium stockpile for an International Nuclear Fuel Bank supply of nuclear fuel for peaceful means under the International Atomic Energy Agency." On August 4, 2008, the U.S. Secretary of Energy issued an official letter to the IAEA donating "nearly 50 million" to the international nuclear fuel bank.[101] This reflects a congressionally mandated rescission that was applied proportionally across the Department of Energy's budget.[102] The IAEA received the U.S. government contribution, which was held in a suspense account until the Board of Governors approved the LEU bank.

Russian LEU Fuel Reserve, Angarsk

The IAEA Board of Governors in November 2009 authorized the director general to sign an agreement with Russia establishing an LEU fuel reserve at Angarsk (GOV/2009/81). The IAEA Director General attended the opening of the reserve in December 2010. The fuel is to be available to a country facing a disruption of supply "unrelated to technical or commercial reasons."[103] The reserve consists of about 120 MT of LEU in the form of UF_6 with an enrichment level ranging from 2 to 4.95% and would be under IAEA safeguards.[104] The Russian government covers the cost of safeguards and all operating costs.

---

[98] Sylvia Westall, "Obama-backed nuclear fuel bank plan stalled at IAEA," *Reuters*, June 18, 2009.

[99] The language is found in H.Rept. 110-477 and was incorporated into P.L. 110-181 by reference.

[100] H.Rept. 110-477 to accompany H.R. 1585. Later incorporated into P.L. 110-181.

[101] "U.S. Donates $50 million for the IAEA International Nuclear Fuel Bank," *NNSA Press Release*, August 4, 2008. http://nnsa.energy.gov/news/2090 htm.

[102] Division C, Title III, Section 312 of the FY2008 Consolidated Appropriations Act rescinded 1.6 percent of discretionary budget authority for Congressionally directed projects, this includes the fuel bank. http://www.whitehouse.gov/omb/legislative/fy08consolidated_reductions_01_25_08.pdf.

[103] "Board of Governors Approves Plan for Nuclear Fuel Bank: Russian Plan to Supply Low-Enriched Uranium," IAEA Staff Report, November 27, 2009, http://www.iaea.org/NewsCenter/News/2009/nuclfuelbank html.

[104] "Development of the initiative of the Russian Federation to establish a reserve of low enriched uranium for the supply of LEU to the International Atomic Energy Agency for its member states," Working Paper of the Russian Federation, Preparatory Committee for the 2010 NPT Review Conference, May 6, 2009, (continued...)

The Russian plan envisions that countries facing a fuel supply cut-off would apply to the IAEA to access the fuel reserve. The director general would assess whether the country meets the criteria for access. To qualify, the potential recipient state would have to be a non-nuclear-weapon state member of the IAEA with a safeguards agreement in effect. To release the material, the IAEA would have to confirm that all nuclear material was accounted for in the state, there was no indication of diversion of material, and no safeguards issues were under review by the Board of Governors. If the criteria were met, the director general would ask Russia to release fuel to that country. The recipient country would pay market rates for the uranium. A country would not have to waive its right to develop its own fuel cycle capabilities in order to access the fuel reserve.

Assurance of Supply: Enrichment Services

Proposals to assure supply and prevent the spread of enrichment technology have also included the creation of commercially based multinational uranium enrichment centers. The Russian Federation has made progress in establishing an International Uranium Enrichment Center (IUEC) at Angarsk. Germany has proposed an internationally owned and operated Multilateral Enrichment Sanctuary Program. Some point out that URENCO, a joint German, Dutch and British consortium, has demonstrated a multilateral commercial model for uranium enrichment since the 1970s. Some countries are concerned that giving support for multilateral-owned facilities would undermine their rights to nuclear technology for peaceful purposes under the NPT, and view the only solution to energy security as being an independent fuel cycle. However, as this may not be economically viable for most countries, multilateral solutions continue to be attractive.

International Uranium Enrichment Center (IUEC), Angarsk, Russia

Russia has established the International Uranium Enrichment Center (IUEC) at Angarsk (approximately 3,000 miles east of Moscow).[105] The Angarsk IUEC began operation on September 5, 2007. Kazakhstan was the first partner,[106] Armenia and Ukraine have since joined. As part of an open joint-stock company, IUEC participants would receive dividends from IUEC profits. Shareholders include Rosatom at 70% of stock, Kazatomprom at 10%, Ukraine's State Concern Nuclear Fuel at 10% and Armenian NPP at 10%.[107]

To join the Angarsk IUEC, countries must agree that the material be used for "nuclear energy production."[108] The IUEC is "chiefly oriented to States not developing uranium enrichment capabilities on their territory."[109] Russia now includes the IUEC on its list of Russian facilities that could be placed under IAEA safeguards.

(...continued)
NPT/CONF.2010/PC.III/WP.25.

[105] Anya Loukianova, "The International Uranium Enrichment Center at Angarsk: A Step Towards Assured Fuel Supply," NTI Issue Brief, updated November 2008, http://www.nti.org/e_research/e3_93.html.

[106] A Russian-Kazakh joint venture Uranium Enrichment Center (distinct from the IUEC) is also located at Angarsk.

[107] "IUEC Corporate Structure," http://eng.iuec.ru/about/corp_structure/.

[108] "Russia's Angarsk international enrichment center open for business," *Nuclear Fuel*, September 24, 2007.

[109] "Communication received from the Resident Representative of the Russian Federation to the IAEA on the Establishment, Structure and Operation of the International Uranium Enrichment Centre," INFCIRC/708, June 8, 2007. Russia, as a nuclear weapon state under the NPT, has a voluntary safeguards agreement that allows, but does not (continued...)

Germany's Multilateral Enrichment Sanctuary Project (MESP)

Germany proposed in May 2007 that a new enrichment facility be built and placed under IAEA ownership in an extraterritorial area.[110] An independent management board or consortium would finance and run the plant on a commercial basis, but the IAEA would decide whether to supply enriched fuel according to nonproliferation criteria. Germany argues that this approach is advantageous since it does not prohibit uranium enrichment, but does provide a commercially viable, politically neutral option for fuel supply and could create competition on the world market by creating a new fuel service provider. With an economically viable option on neutral ground, it will be harder for states to justify starting their own enrichment program for commercial reasons. A proposal was presented to the IAEA Board of Governors in June 2009, and a draft MESP agreement (INFCIRC/765) submitted to the IAEA in July 2009. This concept has not yet been approved by the Board of Governors.

Back-End Fuel Cycle Proposals

Multilateral proposals for the end of the fuel cycle are less developed at this stage. On-site storage of spent fuel is most common, and some countries reprocess their spent fuel rods into mixed-oxide fuel. Multilateral solutions to the back-end issues are also motivated by the idea of preventing the further spread of reprocessing technology, which could be used for the separation of plutonium for weapons purposes. As with the debate over uranium enrichment technology, while reprocessing technology currently may be too expensive or technologically out of reach for many states, countries are hesitant to agree to multilateral approaches that may be interpreted as giving up their right to develop this technology for peaceful purposes. There are no multilateral reprocessing facilities now proposed.

Another proposal has been the establishment of an international spent fuel repository, perhaps in Russia. While Russian law allows for the import of waste, the government of Russia has not yet proposed such a facility, partly due to potential public opposition.

States also cooperate on joint research ventures on advanced and fast reactors such as the Generation IV International Forum (GIF) or IAEA's INPRO. A major U.S.-led initiative, the International Framework for Nuclear Energy Cooperation (IFNEC)—formerly the Global Nuclear Energy Partnership (GNEP)—is intended to foster international collaboration on developing a proliferation-resistant closed fuel cycle, as discussed below.

International Framework for Nuclear Energy Cooperation (IFNEC)

The International Framework for Nuclear Energy Cooperation was initiated in February 2006 by the George W. Bush Administration as the Global Nuclear Energy Partnership (GNEP). Its major purposes were to develop "proliferation resistant" reprocessing technology and to encourage the concentration of reprocessing capacity in a limited number of advanced countries that would agree to provide reprocessing services to any country without such capability. Representatives of

---

(...continued)

require, inspections.

[110] "Safe Enrichment for All," Handelsblatt newspaper, May 2, 2007, in English at https://www.diplo.de/diplo/en/Infoservice/Presse/Interview/2007/070502-Handelsblatt.html.

16 countries signed GNEP's two-page Statement of Principles on September 16, 2007, to launch the organization.

GNEP's domestic activities largely consisted of DOE's Advanced Fuel Cycle Initiative (AFCI), a program that began in 2003 to develop and demonstrate spent fuel reprocessing/recycling technology. However, the Obama Administration rejected AFCI's goal of commercializing advanced reprocessing technology as rapidly as possible. Instead, the program was refocused on fundamental research and development, with similar funding levels, and renamed Fuel Cycle Research and Development.

The international GNEP organization, now with 31 participating countries, changed its name to IFNEC in June 2010 and replaced the original Statement of Principles with a one-paragraph mission statement:

> The International Framework for Nuclear Energy Cooperation provides a forum for cooperation among participating states to explore mutually beneficial approaches to ensure the use of nuclear energy for peaceful purposes proceeds in a manner that is efficient and meets the highest standards of safety, security and non-proliferation. Participating states would not give up any rights and voluntarily engage to share the effort and gain the benefits of economical, peaceful nuclear energy.

According to IFNEC's website, the new mission statement is intended to give the organization "a broader scope with wider international participation to more effectively explore the most important issues underlying the use and expansion of nuclear energy worldwide."[111]

The international component of GNEP as originally envisioned was a consortium of nations with advanced nuclear technology that would provide fuel services and reactors to countries that "refrain" from fuel cycle activities, such as enrichment and reprocessing. It was essentially a fuel leasing approach, wherein the supplier would take responsibility for the final disposition of the spent fuel. This could mean taking back the spent fuel, but might also mean, according to DOE, that the supplier "would retain the responsibility to ensure that the material is secured, safeguarded and disposed of in a manner that meets shared nonproliferation policies."[112]

GNEP foresaw a system whereby supplier states would take back spent fuel, although public opposition to similar proposals in the past has been extremely strong. Some type of "cradle to grave" nuclear fuel management system is still an implicit element of IFNEC, although it is not specifically mentioned in the mission statement. Skeptics of GNEP had questioned whether the reprocessing technology being developed under AFCI would have been a net gain for nonproliferation efforts, since the United States does not reprocess or re-use plutonium now. In their view, the "proliferation-resistance" of technologies under consideration must be assessed against the major alternative: disposal of sealed, intact fuel rods in a geologic repository. At the direction of the White House, Energy Secretary Steven Chu established the Blue Ribbon Commission on America's Nuclear Future on January 29, 2010, to recommend a new national strategy for managing spent nuclear fuel and high-level waste, including an examination of reprocessing and recycling options. The Blue Ribbon Commission issued its final report on January 26, 2012, supporting efforts to develop "spent fuel 'take-away' arrangements," as well as

---

[111] IFNEC website, http://www.ifnec.org.

[112] DOE Global Nuclear Energy Partnership home page, at http://www.gnep.energy.gov.

a research, development, and demonstration program on advanced reactor and fuel cycle technologies.[113]

Much of the research under AFCI had focused on a separations technology called UREX+, in which uranium and other elements are chemically removed from dissolved spent fuel, leaving a mixture of plutonium and other highly radioactive elements. Proponents believe UREX+ is proliferation-resistant, because further purification would be required to make the plutonium useable for weapons and because its high radioactivity would make it difficult to divert or work with. In contrast, conventional reprocessing using the PUREX process can produce weapons-useable plutonium that can be processed in unshielded gloveboxes.

However, critics see the potential nonproliferation benefits of UREX+ over PUREX as minimal. Richard Garwin suggested in testimony to Congress in 2006 that UREX+ fuel fails the proliferation-resistance test. Since it contains 90% plutonium, it could be far more attractive to divert than current spent fuel, which contains 1% plutonium. In other words, a terrorist would have to reprocess only 11 kg of UREX+ fuel to obtain roughly 10 kg of plutonium, in contrast to reprocessing 1,000 kg of highly radioactive spent fuel to get the same amount from light water reactor spent fuel.[114] Under the Obama Administration, the Fuel Cycle Research and Development program is no longer focusing on UREX processes but is instead conducting long-term research that could support a broad range of technology options, according to the DOE FY2011 budget justification.

Another nonproliferation-related concern about GNEP was how its implementation would have affected global stockpiles of separated plutonium. Frank Von Hippel pointed to costly plutonium recycling programs in the United Kingdom, Russia, and Japan, where separated plutonium stocks have accumulated to 320 tons, enough for more than 30,000 nuclear warheads. In Von Hippel's view, GNEP would have exchanged the safer on-site spent fuel storage at reactors for central storage of separated transuranics and high-level waste, cost many times more, and increased the global plutonium stockpile.[115]

A separate set of questions focused on how effective GNEP would have been in achieving its goals. By offering incentives for the back end of the fuel cycle, GNEP was designed to attract states to participate in the fuel supply assurances part of the framework. However, back-end fuel cycle assurances would require significant changes in policies and laws, as well as efforts to commercialize new technologies. Further, it is far from clear that all suppliers would be able to offer the full range of fuel cycle assurances, raising the question of the relative competitiveness of suppliers. Critics did not necessarily argue that the overall vision of GNEP was misplaced, but were generally skeptical that its vision could be achieved, particularly in the timeframe proposed.

---

[113] Blue Ribbon Commission on America's Nuclear Future, *Report to the Secretary of Energy*, January 26, 2012, http://cybercemetery.unt.edu/archive/brc/20120620220235/http://brc.gov/sites/default/files/documents/brc_finalreport_jan2012.pdf.

[114] Richard Garwin, "R&D Priorities for GNEP," Testimony to House Science Committee, April 6, 2006.

[115] Frank von Hippel, "GNEP and the U.S. Spent Fuel Problem," Briefing for Congressional Staff, March 10, 2006, at http://www.princeton.edu/~globsec/publications/pdf/HouseBriefing10March06rev2.pdf; Frank von Hippel, "Managing Spent Fuel in the United States: The Illogic of Reprocessing," International Panel on Fissile Materials, January 2007, at http://www.fissilematerials.org/ipfm/site_down/ipfmresearchreport03.pdf; World Nuclear Association, "Mixed Oxide (MOX) Fuel," May 2012, http://www.world-nuclear.org/info/inf29.html.

GNEP itself marked a departure from a U.S. policy of not encouraging the use of plutonium in civil nuclear fuel cycles. Supporters suggested that the U.S. policy developed in the late 1970s did not envision a recycling process that would not separate pure plutonium, and therefore questioned the underlying assumptions of that longstanding policy. Critics of GNEP have suggested that even though many nations did not agree with the United States in the 1970s on the dangers of having stockpiles of separated plutonium, the message that the United States conveyed was that reprocessing was unnecessary to reap the benefits of nuclear power and that GNEP conveyed the opposite message. Moreover, some critics pointed to the accumulation since the 1970s of separated plutonium as a particular threat, given the potential for terrorist interest in acquiring nuclear material.

An October 2007 study published by the National Research Council recommended that research and development activities for new reprocessing plants continue, but be scaled back, with more time and peer review before commercial plants are built. It strongly criticized DOE's timeline for the program, saying that "achieving GNEP's goals are too early in development to justify DOE's accelerated schedule for construction of commercial facilities that would use these technologies."[116] Congress also expressed significant concerns about GNEP, particularly over the Bush Administration's ambitious schedule for developing fuel cycle demonstration facilities by FY2020.

Under the original GNEP concept, it proved difficult for the United States and others to define which states were suppliers of fuel cycle services and which would be recipients. Informally, U.S. policy currently recognizes 10 states as having enrichment capability—the five nuclear weapon states (United States, United Kingdom, France, China, Russia) plus Japan, Argentina, Brazil, the Netherlands, and Germany. While Argentina has a plant (Pilcaniyeu) under safeguards, this plant has never operated commercially and it is doubtful that it will be cost-effective, since it uses outdated gaseous diffusion technology. Brazil's centrifuge enrichment plant at Resende is still in the early stages of commissioning and won't produce at a commercial scale for several years. States such as Australia, Canada, South Africa, and Ukraine have stated they might consider developing enrichment capability for export in the future. South Korea has also indicated that it would like to build an enrichment plant for its growing nuclear fleet and potentially for fuel exports. On the reprocessing side, South Korea has expressed interest in becoming a GNEP/IFNEC supplier state through development of a pyroprocessing technique that does not separate plutonium from uranium. In the past, the United States, for proliferation reasons, has rejected requests from South Korea to reprocess U.S.-origin spent fuel.

Supply-Side Approaches

Nuclear Suppliers Group

Members of the Nuclear Suppliers Group (NSG), a voluntary group of countries that coordinates nuclear exports and has developed guidelines for such exports, have since the 1970s, adhered to an informal restriction on transferring enrichment, reprocessing, and heavy water technology to

---

[116] "DOE's Spent Nuclear Fuel Reprocessing R&D Program Should Be Scaled Back; Boosted Efforts to Get New Nuclear Power Plants Online Needed," National Academies News Release, October 27, 2007, http://www8.nationalacademies.org/onpinews/newsitem.aspx?RecordID=11998.

states outside the NSG, which currently has 46 members. These policies are voluntary, but resulted in no contractual transfers of enrichment or reprocessing technology to new states.

Following revelations about a covert procurement network for nuclear technology run by former Pakistani nuclear official Abdul Qadeer Khan, some NSG countries sought to tighten these restrictions. NSG member states began in 2004 to negotiate a list of criteria that recipient states would first need to meet before they could receive enrichment or reprocessing technology.[117] In June 2011, after years of debate, the NSG came to a decision on amending Sections 6 and 7 of Part 1 of the Guidelines. The revised guidelines still urge restraint, and add a list of nonproliferation-related criteria for potential recipients.[118]

These criteria require a potential recipient to be an NPT state-party in good standing; to have a comprehensive safeguards agreement in force; to have no current breaches of safeguards obligations; to have a bilateral agreement with the supplier that contains nonproliferation assurances; to commit to international standards of physical protection and safety; and to implement effective export controls and adhere to the NSG guidelines. In addition, the amended guidelines require a recipient state to have brought into force an Additional Protocol to its IAEA safeguards agreement or, "pending this," to implement "appropriate safeguards agreements in cooperation with the IAEA, including a regional accounting and control arrangement for nuclear materials, as approved by the IAEA Board of Governors."

As in the previous version of the guidelines, special provisions are made for transfer of enrichment facilities, equipment, and technology. Suppliers of enrichment plants should also seek agreement from the recipient state that the enrichment facility will not be used to produce greater than 20% enriched uranium. If an enrichment facility was based on a technology that has been demonstrated on a "significant scale" before December 31, 2008, then the supplier should build the plant to prevent the recipient state from replicating the technology transferred (so-called black box transfer). It also specifies that safety and regulatory information should be shared "to the extent necessary without divulging enabling technology."

In negotiations over these revisions, the NSG was also considering the addition of more subjective criteria such as general conditions of stability and security, potential negative impact on the stability and security of the recipient state and the region, and whether there is a credible and coherent rationale for pursuing enrichment and reprocessing capability for civil nuclear power purposes.[119] These proved controversial for a number of states. The June 2011 decision decided to revise the guidelines in this way: "also taking into account at their national discretion, any relevant factors as may be applicable."

Negotiations over the revisions had been contentious. Little public information is available about NSG discussions, but press reports said that Turkey raised objections during the 2010 NSG plenary meeting to several criteria, including the "black box" requirement and subjective criteria

---

[117] Initially, the United States had objected to the criteria-based approach, favoring a moratorium on transfers, until spring 2008 when the Bush Administration changed its policy.

[118] Full text of the revised guidelines are in INFCIRC 254\Rev.10\Part 1, July 26, 2011, http://www.iaea.org/Publications/Documents/Infcircs/2011/infcirc254r10p1.pdf.

[119] See the November 2008 NSG discussion draft, available as Appendix 2 in Fred McGoldrick, *Limiting Transfers of Enrichment and Reprocessing Technology: Issues, Constraints, Options*, Harvard Kennedy School Belfer Center for Science and International Affairs, May 2011. http://belfercenter ksg harvard.edu/files/MTA-NSG-report-color.pdf.

concerning regional stability.[120] In the past, Argentina, Brazil, and South Africa had raised objections to the Additional Protocol as a condition of supply; the provision allowing a "regional accounting and control arrangement" to substitute for an Additional Protocol appears, in effect, to exempt Argentina and Brazil from the Additional Protocol requirement.[121] In general, developing countries are wary of what they characterize as additional obstacles to their ability to access nuclear technology for peaceful purposes.

Some analysts question whether fuel assurance proposals or commercial arrangements would have addressed this issue more effectively without being as contentious as a change in NSG rules.[122] It can also be noted that besides South Korea, no state that does not currently hold the technology is actively seeking to acquire enrichment or reprocessing capability. Additionally, new enrichment plants built in the past few years—by Russia in China, by Urenco and Areva in the United States—have already been based on a "black box" model.

Group of Eight Nations (G-8)

The Group of Eight (G-8) Nations has been the forum where joint policy statements have been made on this issue in recent years. From 2004-2007, the Group of Eight (G-8) Nations announced a year-long suspension of any such transfers at their annual summit meetings. The 2008 Summit declaration stated:

> We agree that transfers of enrichment equipment, facilities and technology to any additional state in the next year will be subject to conditions that, at a minimum, do not permit or enable replication of the facilities; and where technically feasible reprocessing transfers to any additional state will be subject to those same conditions.[123]

Since in 2009 there still had been no agreement on the transfer criteria in the NSG, the G-8 countries said that they would implement this policy on a national basis. The 2010 G-8 Summit statement reaffirmed this commitment:

> To reduce the proliferation risks associated with the spread of enrichment and reprocessing facilities, equipment and technology, we welcome the progress that continues to be made by the Nuclear Suppliers Group (NSG) on mechanisms to strengthen controls on transfers of such enrichment and reprocessing items and technology. While noting that the NSG has not yet reached consensus on this issue, we agree that the NSG discussions have yielded useful and constructive proposals contained in the NSG's "clean text" developed at the 20 November 2008 Consultative Group meeting. Pending completion of work in the NSG, we

---

[120] The U.S. reportedly insisted on including a requirement that uranium enrichment be exported only through a "black box" arrangement. "Black box" or turn-key plants would be built so that recipients would not be able to replicate the facilities, including sensitive components. Canada has reportedly lifted its earlier objections to this provision. Elaine M. Grossman, "Turkish Opposition Delays Deadlock on Proposed Nuclear Trade Guidelines," *Global Security Newswire*, July 2, 2010. For a more recent account of such objections, see Mark Hibbs, "New Global Rules for Sensitive Nuclear Trade," *Nuclear Energy Brief*, Carnegie Endowment for International Peace, July 28, 2011.

[121] Brazil and Argentina formed such a regional arrangement, the Brazilian-Argentine Agency for Accounting and Control of Nuclear Materials, in 1991. However, its provisions are not equivalent to those of an Additional Protocol. For more information, see http://www.abacc.org.br/?page_id=5&lang=en.

[122] Fred McGoldrick, "The Road Ahead for Export Controls: Challenges for the Nuclear Suppliers Group," *Arms Control Today*, January/February 2011.

[123] See paragraph 66 of the Hokkaido Toyako G-8 Summit Leaders Declaration, July 8, 2008, http://www.mofa.go.jp/policy/economy/summit/2008/doc/doc080714__en.html.

agree to implement this text on a national basis in the next year. We urge the NSG to accelerate its work and swiftly reach consensus this year to allow for global implementation of a strengthened mechanism on transfers of enrichment and reprocessing facilities, equipment, and technology.[124]

Comparison of Proposals

Table 4 provides a comparison of the major proposals currently in circulation to restrict sensitive nuclear fuel technology development. The table is based on one created by Chaim Braun presented at the September 2006 IAEA conference on nuclear fuel supply assurances.[125]

---

[124] See paragraph 8 of the L'Aquila G-8 Summit's Statement on Nonproliferation, July 2009, http://www.g8italia2009.it/static/G8_Allegato/2._LAquila_Statent_on_Non_proliferation.pdf.

[125] The IAEA proposal is "Multilateral Approaches to the Nuclear Fuel Cycle: Expert Group Report Submitted to the Director General of the International Atomic Energy Agency," INFCIRC/640, International Atomic Energy Agency, February 22, 2006, p. 18. The Six Country Concept is "Concept for a Multilateral Mechanism for Reliable Access to Nuclear Fuel," Proposal as sent to the IAEA from France, Germany, the Netherlands, Russia, Ireland, and the United States, May 31, 2006. Available at http://www-pub.iaea.org/MTCD/Meetings/PDFplus/2006/cn147_ConceptRA_NF.pdf.

Table 4. Comparison of Major Proposals on Nuclear Fuel Services and Supply Assurances

	IAEA/INFCIRC/640	Putin Initiative	IFNEC	Six Country Concept	World Nuclear Association
Goals	Identify multilateral approaches across the fuel cycle; improve non-proliferation assurances without disrupting market mechanisms.	Establish international commercially operated nuclear fuel service centers in Russia, to include enrichment, education and training, and spent fuel management.	Provide a forum for exploring mutually beneficial approaches to ensuring that nuclear energy expansion proceeds in a manner that is efficient and meets the highest standards of safety, security and non-proliferation.	Create interim measures for front-end assurances.	Enhance supply security.
Target	Front-end and back-end services including uranium enrichment, fuel reprocessing, and disposal and storage of spent fuel.[a]	Supply of nuclear fuel and possibly other fuel cycle services.	Establish international framework for reliable, cost-effective, and proliferation-resistant supply of nuclear fuel cycle services. Facilitate infrastructure for safe, proliferation-resistant expansion of nuclear power.	Supply of nuclear fuel.	Primarily fuel supply.
Methods	Reinforce commercial contracts with transparent supplier arrangements with government backing. International supply guarantees backed by fuel reserves.	Commercial, long-term contracts; recipients will have limited control over joint ventures. IAEA will be involved.[b]	Working group to "recommend critical pathways forward in the development of nuclear fuel service arrangements, including cradle-to-grave fuel management."	Level I: Market Level II: Fuel assurance mechanism at IAEA Level III: Mutual commercial back-up arrangements Level IV: Enriched uranium reserves	Level I: Market meets demand Level II: Standard back-up supply clause in enrichment contracts, with IAEA assurances Level III: Gov't stocks of enriched uranium
IAEA Role	IAEA participates in administering supply guarantees, possibly as guarantor of service supplies with use of a fuel bank. Possible IAEA supervision of an international consortium for reprocessing services.	IAEA would ensure supply with fuel bank created by purchasing existing fuel stocks and placing them under its control (IAEA would receive new funding to do so).	IAEA would apply safeguards.	IAEA as broker. IAEA assesses status of safeguards agreements, safeguards implementation, safety, physical protection and whether a country is pursuing sensitive fuel cycle activities.	IAEA would approve "triggering" mechanism for supply back-up. IAEA could manage enriched uranium reserve.

	IAEA/INFCIRC/640	Putin Initiative	IFNEC	Six Country Concept	World Nuclear Association
Eligibility	Recipient countries would renounce the construction and operation of sensitive fuel cycle facilities and accept safeguards of the highest current standards including comprehensive safeguards and the Additional Protocol.	Equal access, but prerequisite is compliance with the nonproliferation regime. Potential provider states could include Australia and Canada.	Agreement with mission statement (versus initial requirement for recipient states to forgo enrichment and reprocessing).	IAEA-approved states that are in good NPT standing. States that develop national capabilities will not be eligible.	IAEA-approved states that meet all NPT obligations.
Role of Industry	Managing, operating centers.	Performing fuel services at designated center.	Not specified.	Perform enrichment contracts; identified need to address back-end of fuel cycle.	Perform enrichment contracts. No new capabilities required.
Potential Concerns	No mechanism specified for assessing state's nonproliferation record.	Incentives not specified, as well as compliance with nonproliferation regime. Unclear how commitments to forgo sensitive fuel cycle activities will be incorporated into contracts.	Lack of political will to take back spent fuel. Concerns about gains for nonproliferation, if the United States was not reprocessing to begin with.	Incentives may be insufficient.	Incentives may be insufficient. How to determine price on enriched uranium reserves, if they are required.

a. INFCIRC/640, p. 103.

b. "Questions Abound on Proposals by Bush, Putin on Fuel Centers," *Nuclear Fuel*, March 13, 2006, vol. 31, no. 3.

Prospects for Implementing Fuel Assurance Mechanisms

Proposals to provide an international and institutional framework for peaceful nuclear activities have abounded since the 1940s, but few have been implemented. The U.S.-sponsored Baruch Plan introduced at the United Nations in 1946 recommended establishing an international agency with managerial control or ownership of all atomic energy activities. The International Atomic Energy Agency, established in 1957, emerged as a paler version of what was suggested in the Baruch Plan, but still retains authorities in its statute to store fissile material.

Concern about proliferation led to a flurry of proposals in the 1970s and 1980s as the United States and others convened groups to study the issues.[126] One idea studied in the mid-1970s was regional nuclear fuel cycle centers, focused on reprocessing technologies. Several factors contributed to its lack of success, despite support by the U.S. Congress: low uranium prices (making plutonium recovery relatively unattractive), a slump in the nuclear industry in the late 1970s and early 1980s, and U.S. opposition to reprocessing from the late 1970s. Member states of the IAEA also convened the International Fuel Cycle Evaluation (INFCE) project, which involved 60 countries and international organizations. INFCE working group reports suggested establishing a multi-tiered assurance of supply mechanism similar to the one proposed by the Six Country Concept in 2006. States also studied international plutonium storage in the late 1970s and early 1980s, but could not agree on how to define excess material or the requirements for releasing materials.

As in the past, the success of current proposals may depend on whether nuclear energy is truly revived not just in the United States, but globally. That revival will likely depend on significant support for nuclear energy in the form of policy, price supports, and incentives. Factors that may help improve the position of nuclear energy against alternative sources of electricity include higher prices for other sources (natural gas and coal through a carbon tax or other restrictions), improved reactor designs to reduce capital costs, regulatory improvements, and waste disposal solutions.

The willingness of fuel recipient states to participate in international enrichment centers rather than develop indigenous enrichment capabilities, and confidence in fuel supply assurance mechanisms such as an international fuel bank, will largely determine the success of the overall policy goal—to prevent further spread of enrichment and reprocessing technologies. So far, proposals addressing this challenge have originated in the supplier states, with many recipient states continuing to voice concern that their right to peaceful nuclear energy technology under the NPT is in jeopardy. Increasingly, however, participation is being presented as a market-based decision by countries to refrain, at least for the present, from developing their own fuel enrichment programs.

Another factor that will shape the success of these proposals is the possible addition of other incentives. Simply making nuclear energy cost-effective may not induce countries to forgo indigenous enrichment and reprocessing. Such decisions may require other incentives, perhaps

---

[126] For an analysis of these past proposals, see Lawrence Scheinman, "Equal Opportunity: Historical Challenges and Future Prospects of the Nuclear Fuel Cycle," *Arms Control Today*, May 2007.

even outside the nuclear realm, to make them palatable. The experience of Iran may be instructive here. Russia's offer to provide assured enrichment services on Russian soil has gone nowhere; instead, other, broader trade incentives may be necessary. While the case of Iran may illustrate the extreme end of the spectrum, in terms of a country determined to develop a capability for a weapons program, non-nuclear-weapon states will clearly take notice of how a solution develops for Iran.

Issues for Congress

Congress would have a considerable role in at least four areas of oversight related to fuel cycle proposals. The first is providing funding and oversight of U.S. domestic programs related to expanding nuclear energy in the United States. Key among these programs are nuclear research and development programs and federal incentives for building new commercial reactors.[127]

The second area is policy direction and/or funding for international measures to assure supply. What guarantees should the United States insist upon in exchange for helping provide fuel assurances? Although the Six Country Concept contains an option for a fuel bank, it would not require participants to forswear enrichment and reprocessing.

A third set of policy issues may arise in the context of development of the International Framework for Nuclear Energy Cooperation. Observers may question what the nonproliferation benefits of this program are, how it overlaps with other programs such as those under the IAEA, and what the United States aims to achieve through IFNEC. The new mission statement emphasizes that members do not give up any rights under the NPT to the peaceful use of nuclear energy. Policymakers may explore whether the newly envisioned program goes far enough in encouraging states to refrain from enrichment and reprocessing, a key goal of the original international GNEP.

Some observers believe that further restrictions on non-nuclear-weapon states party to the NPT are untenable in the absence of substantial disarmament commitments by nuclear weapon states. In particular, a January 4, 2007, *Wall Street Journal* op-ed by George Shultz, Bill Perry, Henry Kissinger, and Sam Nunn, entitled "A World Free of Nuclear Weapons," noted that non-nuclear-weapon states have grown increasingly skeptical of the sincerity of nuclear weapon states in this regard. Some observers have asserted that non-nuclear-weapon states will not tolerate limits on NPT Article IV rights (right to pursue peaceful uses of nuclear energy) without progress under Article VI of the NPT (disarmament). Amending the NPT is seen by most observers as unattainable. President Obama called for the eventual elimination of nuclear weapons in a speech in the Czech Republic on April 5, 2009.

The IAEA experts group report, INFCIRC/640, did point to the political usefulness of achieving a ban on producing fissile material for nuclear weapons (known as fissile material production cutoff treaty, or FMCT) to provide more balance between the obligations of nuclear and non-nuclear-weapon states. Obama Administration officials have indicated that they will pursue negotiations on a fissile material cut-off treaty that includes verification provisions. Ultimately, any such treaty would require Senate advice and consent to ratification.

---

[127] See CRS Report RL33558, *Nuclear Energy Policy*, by Mark Holt.

A fourth area in which Congress plays a key role is with the approval of nuclear cooperation agreements.[128] In some cases, the United States may seek additional reassurances regarding fuel cycle facilities during negotiations of civilian nuclear cooperation agreements. This was a topic of controversy during the approval process for the civilian nuclear cooperation agreement with India in September 2008.[129] A civilian cooperation agreement with the United Arab Emirates was preceded by a signed a memorandum of understanding with the United States saying it would forgo "domestic enrichment and reprocessing capabilities in favor of long-term commitments of the secure external supply of nuclear fuel." In addition, the nuclear cooperation agreement's text itself states that the United States can end nuclear cooperation with the UAE if it acquires enrichment or reprocessing facilities. Some Members of Congress introduced legislation (H.R. 1280) in the 112th Congress that would amend the Atomic Energy Act to require this commitment in all nuclear cooperation agreements.

---

[128] See CRS Report RS22937, *Nuclear Cooperation with Other Countries: A Primer*, by Paul K. Kerr and Mary Beth Nikitin.

[129] CRS Report RL33016, *U.S. Nuclear Cooperation with India: Issues for Congress*, by Paul K. Kerr.

Figure 2. World Wide Nuclear Power Plants Operating, Under Construction, and Planned

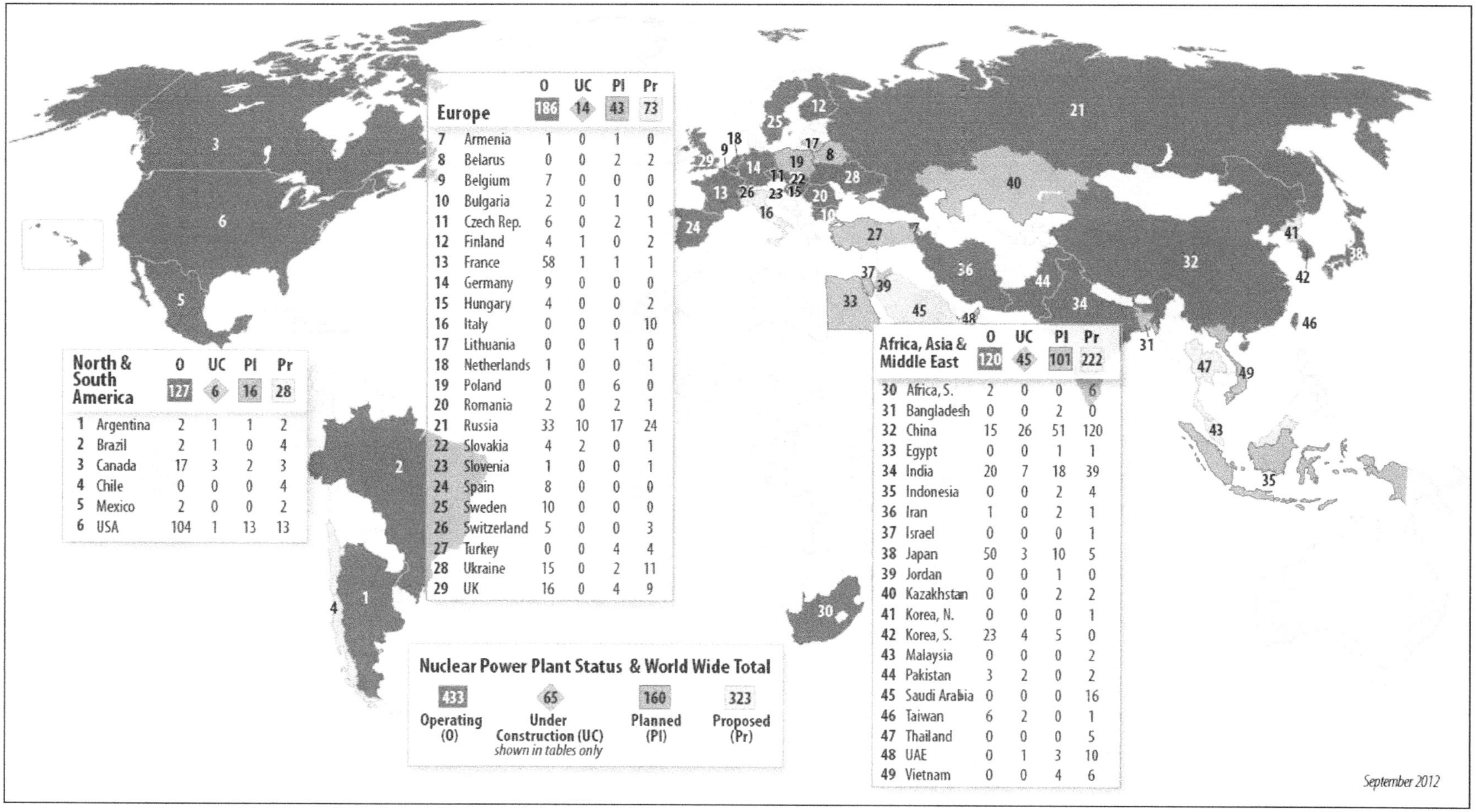

Europe	O 186	UC 14	Pl 43	Pr 73
7 Armenia	1	0	1	0
8 Belarus	0	0	2	2
9 Belgium	7	0	0	0
10 Bulgaria	2	0	1	0
11 Czech Rep.	6	0	2	1
12 Finland	4	1	0	2
13 France	58	1	1	1
14 Germany	9	0	0	0
15 Hungary	4	0	0	2
16 Italy	0	0	0	10
17 Lithuania	0	0	1	0
18 Netherlands	1	0	0	1
19 Poland	0	0	6	0
20 Romania	2	0	2	1
21 Russia	33	10	17	24
22 Slovakia	4	2	0	1
23 Slovenia	1	0	0	1
24 Spain	8	0	0	0
25 Sweden	10	0	0	0
26 Switzerland	5	0	0	3
27 Turkey	0	0	4	4
28 Ukraine	15	0	2	11
29 UK	16	0	4	9

Africa, Asia & Middle East	O 120	UC 45	Pl 101	Pr 222
30 Africa, S.	2	0	0	6
31 Bangladesh	0	0	2	0
32 China	15	26	51	120
33 Egypt	0	0	1	1
34 India	20	7	18	39
35 Indonesia	0	0	2	4
36 Iran	1	0	2	1
37 Israel	0	0	0	1
38 Japan	50	3	10	5
39 Jordan	0	0	1	0
40 Kazakhstan	0	0	2	2
41 Korea, N.	0	0	0	1
42 Korea, S.	23	4	5	0
43 Malaysia	0	0	0	2
44 Pakistan	3	2	0	2
45 Saudi Arabia	0	0	0	16
46 Taiwan	6	2	0	1
47 Thailand	0	0	0	5
48 UAE	0	1	3	10
49 Vietnam	0	0	4	6

North & South America	O 127	UC 6	Pl 16	Pr 28
1 Argentina	2	1	1	2
2 Brazil	2	1	0	4
3 Canada	17	3	2	3
4 Chile	0	0	0	4
5 Mexico	2	0	0	2
6 USA	104	1	13	13

Nuclear Power Plant Status & World Wide Total			
Operating (O) 433	Under Construction (UC) 65 shown in tables only	Planned (Pl) 160	Proposed (Pr) 323

September 2012

Source: World Nuclear Association, http://www.world-nuclear.org/info/reactors.html.

Author Contact Information

Mary Beth Nikitin, Coordinator
Specialist in Nonproliferation
mnikitin@crs.loc.gov, 7-7745

Mark Holt
Specialist in Energy Policy
mholt@crs.loc.gov, 7-1704

Anthony Andrews
Specialist in Energy and Defense Policy
aandrews@crs.loc.gov, 7-6843

Acknowledgments

Jill Marie Parillo and Sharon Squassoni were original contributors to this report.

www.ingramcontent.com/pod-product-compliance
Lightning Source LLC
Chambersburg PA
CBHW081359170526
45166CB00010B/3149